Y0-BOE-853

DEADLY DANCE

Books by Bv Lawson

Scott Drayco Series

Novels

Played to Death
Requiem for Innocence
Dies Irae
Elegy in Scarlet
The Suicide Sonata
Deadly Dance

Short Story Collections

False Shadows
Hear No Evil
Vengeance is Blind

Adam Dutton & Beverly Laborde Series

Steal Away
Hide Away
Burn Away

Deadly Dance

A Scott Drayco Mystery

BV Lawson

Crimetime Press

Deadly Dance is a work of fiction. All of the names, characters, places, organizations and events portrayed in this novel are either products of the author's imagination or are used fictitiously. Any resemblance to actual events, locales, or persons, living or dead, is entirely coincidental.

Published in the United States of America.

For information, contact:

Crimetime Press
6312 Seven Corners Center, Box 257
Falls Church, VA 22044

Trade Paperback ISBN 978-1-951752-07-1
Hardcover ISBN 978-1-951752-08-8
eBook ISBN 978-1-951752-06-4

Seek Not to Know

God of dreams:
Seek not to know what must not be reveal'd,
Joys only flow when hate is most conceal'd.
Too busy man would find his sorrows more
If future fortunes he should know before;

For by that knowledge of his destiny
He would not live at all but always die.
Enquire not then who shall from bonds be freed,
Who'tis shall wear a crown and who shall bleed.

All must submit to their appointed doom,
Fate and misfortune will too quickly come.
Let me no more with powerful charms be press'd
I am forbid by fate to tell the rest.

poem by John Dryden, music by Henry Purcell

1

Tuesday, September 15

The first sign of trouble came at eleven when Scott Drayco's cellphone rang. Attorney Benny Baskin's voice chirped in his ear, "Just giving you a heads-up, boy-o. I'm sending over someone who wants to talk with you."

"Who?"

"A client of mine. You'll see."

"Can't you give me any more details than that?"

"I predict you'll have a sudden desire to stop by my office afterward. Say tomorrow, about this same time?"

"Benny, what in the world—"

"Oops, there's another incoming call. Gotta go."

The attorney hung up, leaving Drayco with a surprising uneasiness to go along with his unanswered questions. Maybe he should blame it on a friend who'd sent a teasing text with the horoscope for Sagittarians, *Today's a good day to prepare for the unexpected.*

He shrugged it off, tossed the phone aside, and slid onto his piano bench to tackle an engrossing Beethoven sonata. He'd only managed a couple of minutes of the first movement when the doorbell rang. Yet, even knowing someone was coming over thanks to Benny's warning, Drayco almost jumped out of his skin. With a sigh, he gave the shiny Steinway a little pat and headed toward the front door.

When he opened it, a figure stood framed against a backdrop of rain so thick, he could barely see the sidewalk across the street. It was the last person he'd expected after Benny's call—but he now realized why Benny was so cagey on the phone. The universal karmic gods certainly had a very dark sense of humor, and the score was horoscope, one, Scott Drayco, zero.

Darcie Squier pushed her way in, shook the rain off her red umbrella, and slid it into the stand by the door. He had to grudgingly admit she looked as good as ever, wearing a form-fitting turquoise knit dress that set off her dark hair nicely.

What the hell could she want? And what did Baskin have to do with it? When she'd told Drayco about her engagement months ago, Darcie said she still craved the physical relationship they'd shared. Was that why she was here?

He started to formulate a polite, but firm, "No," when she shocked him. "You have to help him. Please say you will. I don't trust anyone else."

Her eyes were moist and not from the rain. Drayco helped her wriggle out of her raincoat and guided her into the den where he settled her on the sofa, heading to the kitchen long enough to make a cup of hot Earl Grey tea he handed over.

"What's this all about, Darcie? Who do you want me to help?" He parked on the chair across from her.

"It's Harry."

"Harry Dickerman? Your fiancé?"

She gulped some of the aromatic tea before replying. "He's been arrested and is in jail. But he didn't do it."

"Arrested on what charge?"

"Suspicion of murder."

Drayco leaned back in his chair. "You've retained Benny Baskin."

"I heard you mention him before. That he was the best. So I hired him, and now he wants you to assist. I'm begging you to say yes, even though I'm afraid you won't."

Drayco had a sudden craving for some tea himself, with a shot of vodka. Or maybe just the vodka. "Why don't you start from the beginning. Tell me what happened."

"The murder was four days ago at our house in McLean. I wasn't there at the time, but this woman arrived saying she needed to see Harry, and she's apparently Harry's ex-wife, well fake ex-wife, and—"

Drayco cut off her rambling. "Fake ex-wife?"

"I think so. I'm not clear on all the details. I'm still in shock. They let me see Harry, but he doesn't know what's going on, either."

"Where were you at the time?"

"At the salon. With lots of witnesses. I came home to find the police there, the body being carted away, and Harry hauled off to jail. So, yesterday I hired Benny Baskin."

"Why wait three days?"

"I didn't know what to do. I've been in such a panic. I wanted to call you."

"But you were afraid I'd give you the brush-off, is that it?"

She nodded.

"Oh, Darcie." Maybe they hadn't shared a deep love, but he'd thought she knew him better than that. "Are you sure Harry will want me involved?"

"I don't care. It's for his own good. He'll just have to understand that."

Drayco rubbed his temples. "All right, then. Assuming Harry didn't murder his ex-wife, then someone might be trying to frame him, but who and why?"

"I have no idea. I haven't known him all that long."

"So I noticed."

"Don't be angry with me, Scott. Not now." She set the tea down and clutched her hands together. "I'm afraid I'll be charged, too, as an accessory or some such thing. That's what they do in the movies."

"Sounds like you have an airtight alibi. That should help."

She was shivering, and that's when he realized how wet her clothing was from the rain. "You're soaked. Go get some of my clothes and give me your things. I'll put them in the dryer." He added quickly, "And you can hang out in the guest bedroom, watch the TV, while I make some calls. All nice and proper."

She gave him a small smile. "Just hang out in the guest bedroom?"

"It's much safer that way, don't you think?"

"I suppose so."

"I'm assuming you aren't staying at Harry's place?"

"A friend's letting me bunk with her until I move back to Cape Unity. I don't feel right being in Harry's house without Harry. Not that I'd want to until the crime-scene people have taken care of the. . .of all of that blood."

Drayco ushered her to the guest bedroom and dutifully took the wet clothes she handed him to the dryer. Truth be told, he was in a bit of shock, himself. He'd have been far more prepared for this if he'd heard of Harry's arrest when it happened. But he was too busy wrapping up a case in Pennsylvania and out of touch with local Washington-area news.

He perched on the chair in his study to make some of those calls, but couldn't stop obsessing over the fake-ex-wife part of Darcie's tale. No doubt, Benny Baskin would fill him in on all the details tomorrow. But a "fake" ex-wife hinted at something deeper and more mysterious than simple jealousy. Just then, a loud clap of thunder rattled the windows. Seems the universe agreed with him.

He gave a longing glance over at the piano, wanting nothing more than to return to the Beethoven Appassionata sonata he was playing before he got Benny's call. Why did stormy weather make him want to play something in a minor key? Must be his synesthesia. Pieces like the sonata with its F minor key created shimmering, elongated sepia ovals—like twisted noirish raindrops—to his ears. Later, he'd switch to something more upbeat, Chopin's Mazurka in D major, maybe. Something, anything to chase away his dark mood.

Right now, he had an ex-lover in his bedroom and a date with the phone to make calls to everyone who might have more intel on the murder case. And to plot revenge against a certain pint-sized attorney who'd arranged all this and left out the details as a little "surprise." Damn the man.

Drayco had the feeling Benny's horoscope today was something along the lines of, *Someone close to you will wish you harm*. Or perhaps that was Harry Dickerman's horoscope four days ago. But exactly who would wish Dickerman harm? As far as Drayco knew, he was a typical boring, rich businessman. . .one with a taste for much younger women.

With that unpleasant thought, Drayco grabbed the phone and started to dial.

2

Wednesday, September 16

Having Darcie in his townhome again was both easier and harder than Drayco expected. It wasn't difficult at all to fall back into the protective lover mode, yet it also sent him down a slope of conflicted emotions he didn't have the time or energy to psychoanalyze.

He certainly hadn't slept well after seeing her off and spending much of the evening digging up everything he could on Harry Dickerman. He found it singularly unhelpful for the most part—scores of career accolades and no hints of scandal. Hopefully, his late-morning meeting would be the eye-opener he needed.

Knowing Benny Baskin's obsession with punctuality, Drayco made sure he was at the attorney's office at fifteen minutes past eleven. Tweaking Benny was one of Drayco's favorite pastimes, although it wasn't easy to take a guy down a peg when he was only four-nine in his platform shoes. He got close once when he gave Benny an eye patch with dollar signs on it to use instead of his customary black one.

All thoughts of the diminutive attorney flew out of Drayco's mind when he bumped into a familiar figure in the hallway, a tall blonde whose hair was pulled back into her usual French braid. He sucked in air through his teeth but managed to paste on a smile. "Fancy meeting you here, Nelia." Two different days, two very different women, and he didn't know how to feel about either of them.

Even though seeing Nelia sent his pulse racing, he'd kept his voice light and joking, and she chose to do the same, as she replied, "Wasn't as if I had anything else to do."

That made him grimace. "I honestly don't see how you go to law school part-time, manage a deputy job part-time, and still intern with Baskin."

"Just call me Super Woman." She drew a big "S" across her chest.

"Can't you get a grant or loan? You could quit the deputy gig and focus on the legal side."

"Not enough grants to go around."

"There must be something—"

"Tim's attorney biz is down due to his illness. We need my part-time salary. This was the only way Tim would agree to me going to law school."

"Doesn't he know what this is doing to you? The toll it's taking on *your* health? And after all you've sacrificed for him?" That came out a little on the harsh side, but Drayco really didn't care.

He almost saw her counting to ten before she replied, "He has his own problems. The MS and all." The usual coppery shimmer to her voice like tinkling wind chimes—which to a synesthete like Drayco was audio honey—turned sharper, forming jagged crystals with brown edges.

He asked, "Does he still have his personal aide helping out?"

"Melanie's the only thing standing between me and insanity. Plus, she seems to know how to handle his moods."

Drayco bit back an even harsher retort since he'd witnessed the violence and cruelty from some of those moods. Instead, he opted to keep it light. "Well, considering your schedule, your caffeine budget must dwarf mine."

She held out her arm and shook it. "I'm getting injections now. Saves time."

He opened the door to Benny's office for her, and they headed on in. As Drayco looked around, the floor-to-ceiling bookshelves and the mahogany coat rack in the corner suddenly struck him as dated. The room reeked of dusty books, old leather, and fake-lemon furniture polish. "Benny, you need to hire Nelia when she gets her J.D. to class up the place."

Benny wagged his finger at him. "She'll have her pick of the best. Not sure I'm enough of a draw." At least Baskin's voice still sounded the same—twice as deep as his stature and a little like a salmon-colored tumbleweed.

Nelia smiled at Baskin. "You'll always be my favorite attorney."

Drayco rolled his eyes. "This mutual admiration society is getting syrupy. How about some murder to spice things up?"

Nelia slid into a padded wing chair while Benny grabbed his notes and perched precariously on the edge of his desk. His feet didn't quite reach the parquet floor.

Drayco stood next to a large globe on a stand as tall as the attorney. "Talk to me, Benny. Convince me of why I should help you on this case."

"Harry Dickerman is President and CEO of Mediasio. It's an international radio and billboard advertising business. He's well-respected, beloved, an upstanding citizen with a record as clean as a bleached sheet."

"Until now."

"He's been charged with the suspicion of murdering his ex-wife, Lara Ekaterina Davidenko, whom he thought he'd divorced fifteen years ago. The Fairfax police are in charge since the murder was in McLean."

"What do you mean he *thought* he'd divorced her?"

"She was a Russian immigrant he took pity on and married for green-card status. Vanished soon after the marriage, and he hadn't heard from her since. Apparently, there was a shifty attorney involved. Supposedly filed the Affidavit of Diligent Search for a divorce in absentia. But I guess he forgot." Baskin smirked.

Drayco mulled that over. "Where has this woman been all this time? Did the Fairfax PD trace her history?"

"Off the grid, they said."

"That's kind of important, don't you think? She's a ghost for the last fifteen years, and all of a sudden turns up at her ex-husband's place as a corpse?"

"Odd, for sure. But odd's right up your alley, boy-o."

"How did the murder happen? Who called the police?"

"Harry, after he found Lara's body in his living room."

Drayco was getting irritated at Benny's roundabout account. But when he saw how much the other man was enjoying Drayco's irritation, he gave in. He probably deserved that. "How did the victim get to Harry's house?"

"She had a leased car and listed an address up in Salisbury. Paid for the lease in cash."

"Cash?"

"Yep, but here's the real fun part. She shows up out of the blue, Harry lets her in. She says she has something she needs to discuss with him, she appears distraught, so he offers to get her some coffee. He goes to the kitchen, isn't gone for much more than a few minutes, returns, and boom. She's dying on the floor, blood everywhere, with his letter opener plunged into her back. And only his prints on it."

"Multiple stab wounds?"

"One, but it was deep. Punctured the pericardium, diaphragm, and liver."

"Let me guess. The police think Harry killed the victim because he was getting ready to marry Darcie—and this woman was in his way."

Benny tapped the side of his nose, just missing his black eye patch. "And then, there's the will."

"The victim's?"

"She made a will a year ago with Harry named the sole beneficiary."

"A year ago? As if this case wasn't bizarre enough." Drayco spun the globe around. "Since Harry is already wealthy, her money seems like a weak motive."

"It just so happens I know the attorney who made up the victim's will."

"Oh, really? Okay, then how much money are we talking about?"

"My attorney-friend couldn't say, of course, but hinted it wasn't much. I did get him to say the victim seemed skittish and a bit afraid. Not ill or anything."

"Was her behavior odd in any other way?"

"She said it was time for her to do this. I think her words were, 'you never know when you were going to suffer a horrible accident.'"

Nelia spoke up, "Seems rather prescient. As if she knew she was in danger."

Drayco frowned. "But to make Harry the beneficiary after fifteen years? And why did she pick him for her green card back then, anyway? And why that particular time? None of this makes much sense."

Benny huffed, "If it helps you feel any better, Harry Dickerman is quite the charitable fellow. On various boards of foundations. Gives a bunch of money to worthwhile causes. Orphans, medical research, literacy, arts organizations."

"I'll start calling him Saint Harry. He's a paragon of virtue who's accused of killing his fake ex-wife to pave the way for marrying Darcie—even when a real divorce would take care of that—and to get his hands on the ex's money that he doesn't need. Does that sum it up?"

Benny replied cheerfully, "Yep. Fun, eh?"

"Your idea of fun is karaoke at the Blue Hayes Lounge. I don't like your definition."

"Why don't we mosey over to the scene of the crime and take a look around for ourselves? The police have cleared it. Maybe it will help pique your interest."

"You mean right now?"

"Sure, why not? I'll tell my secretary to reschedule my eleven o'clock."

"You don't have a secretary."

"My answering service, then." Benny's cheerful tone dipped a bit at that. He hadn't got over his previous secretary "leaving him" for another law firm. Even though it was because her husband took a job in New York.

Nelia added, "You boys want company? I've seen a few crime scenes, myself."

Drayco's interest in the case suddenly took an upturn. "How could I possibly refuse the best CSI agent I've ever met?"

Her eyes danced. "Flatterer. I'll bet you say that to all the crime scene investigators."

"Never. I am not a two-timing detective." He immediately regretted his choice of words as her cheeks flushed a bright pink.

As if sensing the awkwardness crackling through the room, Benny hopped up from his seat and said, "Come on, kids, let's go play shamus. They do still use that term, don't they? And Drayco, you're driving. Especially if you have that Starfire of yours."

"It's the Generic Silver Camry. Sorry."

"Oh." Benny's cheeks puffed out with disappointment. "Guess I'll cope. But you owe me a ride in that classic car of yours. Out on I-66 west of town, where we can crank it up to almost-legal speeds."

"Almost legal, Benny?"

"Hey, you're the one who'd get the ticket, not me. But that's okay. I know a good attorney."

Drayco groaned. He was having second thoughts about the trip if Baskin was going to be popping bad jokes all day. When they reached the car, Nelia quickly took a backseat as if wanting to put as much distance between her and Drayco as possible.

Benny slid into the shotgun seat with a grin. "I hope you remember how to drive. Just be sure not to 'dicker,' man. Get it? Harry Dicker-man?"

Drayco shook his head. Of all the attorneys for him to be associated with, it had to be a wisecracking esquire. He looked into the rearview mirror expecting to see Nelia smiling at the awful pun, but she was staring out the window with a deep frown.

He tried to cheer her up. "I could stop by the convenience mart and pick up some lottery tickets. They say money buys happiness. Or so I'm told."

She was still frowning, but a little less so, as she replied, "My horoscope said, 'Don't fixate on things outside yourself to solve all your problems.'"

He groaned inwardly. Horoscopes, two, Scott Drayco, zero.

3

Drayco had always considered the Eastern Shore home of Darcie Squier and her ex-husband to be a bit on the pretentious side. Even the name, Cypress Manor, was over the top. But that place looked like a squatter's cabin next to Harry Dickerman's McLean residence.

A fountain as big as Drayco's car, topped with a statue of the Greek god, Zeus, loomed in front of the mini-palace made of gleaming white stone and brick. The grounds were as expansive as some museums, and he had the passing thought he'd need to pay admission to enter.

Benny, Nelia, and Drayco navigated up the dozen or so stairs to the front door that Benny opened using a key Darcie had given him. The foyer—or "lobby," which was more like it—was as gleaming white as the exterior and sported vaulted ceilings and a winding staircase leading to the second floor. Drayco listened for a moment. Quiet and still, save for the low humming of the air conditioner.

He led the way into the living room, where the trio stopped and stared at a section of the beige carpet next to some recessed bookshelves. He didn't have to ask Benny if that was where the victim was stabbed, as the dark red stains were a clear giveaway.

Nelia pulled out some nitrile gloves from her pocket and handed some to Drayco as she said pointedly to Benny, "The police didn't find unusual fingerprints or blood. But don't touch anything. Just in case."

"Are you kidding? I hate wearing gloves."

Drayco added, "And if you do touch anything, we'll hold you down and put on some of these gloves by force."

"You wouldn't dare."

"Try me."

Drayco studied the bloodstains. "Shame they didn't recover any DNA evidence of our killer."

Nelia added, "Be nice to try phenotyping."

Benny asked, "What's that?"

"It can narrow the suspect's gender," she replied. "And appearance. It's a fairly new forensic tool."

"Ah, yes. I've heard of that. Might want to use it in a defense, myself, one of these days." Benny waddled over to the bookshelves, stopping short of the stains. "Guess the cleanup crew hasn't been here yet."

Drayco had made Benny read from the police notes about the murder on the way over, and Drayco now pointed to the door-less study off the main room. "Is that where the letter opener used on the victim was obtained?"

"Allegedly."

Drayco said over his shoulder as he headed toward the room, "If the killer needed a weapon, that means they didn't bring one with them. So possibly not premeditation."

Nelia asked, "Or, the killer *did* bring a weapon but saw the opportunity to frame Harry using something from Harry's own home. I mean, it's definitely his, with his initials on it and only his fingerprints, right?"

"So says Benny's copy of the police report." Drayco stood in the arched doorway of the study and used his cellphone to take photos.

Next, he strolled over to the desk and pulled out the drawers. Paper, pens, blank envelopes, and paper clips, very ordinary and disappointingly boring. "Harry says he thinks he left the letter opener lying on top of the desk?"

Benny eyed the desk. "Allegedly. If true, you could see it from the living room. Grab letter opener, stab victim, easy-peasy."

Nelia said, "Except for one thing. If he came through the front door, he would have been seen. Surely both the victim and Harry would have noticed?"

Drayco agreed. "There's got to be another entry point." He popped out into the main living area and peered through a doorway toward the rear of that room. "I think I see the kitchen, way back there in a different zip code. With Harry making drinks, he might have created too much noise to hear an intruder. And he is in his mid-sixties. His hearing's not necessarily top-notch."

Benny growled at him. "Mid-sixties isn't all that ancient, boy-o. You'll be there before you know it. How old are you again? Thirty-six going on a hundred? I keep forgetting."

"Thirty-seven."

"A babe in arms. Good thing you're precocious."

Drayco ignored him. "The police report indicates a neighbor saw no other visitors that night, just the victim. At least in front of the house." He peered down a hallway leading away from the study. "Aha."

"What?"

"A side door with an entryway. I think they call those a 'mud room'" Drayco headed toward the door as his two companions watched from the hall. He called back to them, "It would be hard for someone to jigger the lock without triggering Harry's alarm system."

Benny said, "Did I not read you that part of the report? Harry had a visitor earlier in the day, a friend bringing over some Petit Verdot from a Virginia winery. The alarm wasn't turned back on."

Drayco thought about that for a moment. "A killer wouldn't know that. But if the alarm didn't go off when they tried the door, maybe they figured they were in the clear."

Nelia added helpfully, "Or they didn't see the alarm system since they entered from the back. Or were in a hurry."

"True. No dogs, right?"

Benny shook his head. "Nope."

"Okay, then. Our killer *could* have entered through this side door and surprised the victim enough to stab her in the back."

Drayco headed toward the bloodstains in the living room. "From the blood spatter, it appears the victim was standing right there," he pointed at one particular spot, "in front of that section of bookshelves." He took more cellphone snaps.

Bits of blood dotted a shelf sporting an empty slot—the same width as a book lying on the floor nearby. He added, "Either she was reading that specific book or grabbed onto the shelves when she was stabbed, and the book fell out."

Benny glanced at the bookshelves. "From the looks of it, Harry alphabetizes his books by title. For what it's worth."

Drayco picked up the book from the carpet. "It's a reference of criminal codes." He flipped through it and stopped. "Interesting. There's one torn page. A section on the penalties for murder and accessory to murder."

Nelia peered over Drayco's shoulder. "That's a pretty big coincidence. This is the one that falls to the floor? Out of all these books?" She read off a few other titles from that section. "*Canterbury Tales . . . Chess for Beginners . . . Collecting World Coins . . .*"

Drayco turned to Benny. "You think Darcie or Harry would mind if I kept this book?"

"I doubt it. If it'll help."

Nelia frowned. "I've been wondering how the killer knew the victim would be here at the exact time. Must have stalked her."

"Or tapped her phone. Pretty easy to track people in the digital age." Drayco looked toward the side door. "As I recall from online aerial surveys, there's another street running behind this house."

Nelia replied, "The perp could have arrived that way unseen by the neighbor. Is there a fence in back?"

"According to the aerial maps I consulted, yes, if it's still there." Drayco led the way out the side door toward the rear yard. It was pretty much as Darcie once described to him, with "a giant pool and this huge garden with a grape arbor. Like out of *Architectural Digest*."

Drayco was particularly interested in the dense vegetation of boxwood and barberry bushes. Lots of hiding room, there. He headed toward the pool first, looking around as he asked Benny, "Did the Fairfax PD check for shoeprints here or signs someone climbed the fence? Or blood?"

Benny said, "No, no, and no. Were you sleeping through my reading of the report?"

"You'd put anyone to sleep, Benny." Drayco pointed out a tree behind Harry's fence, an oak around twenty feet tall, with branches hanging over Harry's property. "If you were strong enough and agile, you might be able to climb up the tree, swing into Harry's yard, and then do it in reverse to make your escape."

Benny pursed his lips. "But wouldn't there be a trail of blood all over the place?"

"The letter opener was left in the body, and the police report said there was no arterial spurting. So there might not have been much transfer of blood from the victim to the killer." Drayco pointed to the ground. "And that landscaping mulch means less chance of footprints."

Nelia seemed to agree with Drayco's assessment of the blood. "Blood spatter doesn't always travel far. Since the only fingerprints on the knife were Harry's, our killer likely wore gloves."

Benny eyed the oak skeptically. "There wasn't anything about that tree in the police report."

"There wouldn't be if they were firmly convinced Harry was their man." Drayco gauged how easy it would be for him to haul his six-four frame onto the first tree branch. Never knew until you tried.

As Nelia called out a worried-sounding, "Be careful," he crouched down a few inches and then jumped up, grabbing the top of the fence as he leaped. He managed to hook a leg over a fence post and did his best acrobat impression to swing over to the closest branch.

When Benny started clapping, Drayco said, "Fifteen seconds."

"What?"

"That took me around fifteen seconds. If I can do it, so could someone else."

"Fair enough. Since you're up there, see anything strange?"

Drayco examined the rough, scratchy branch he was perched on. Nothing out of the ordinary. But when he studied the trunk, something that seemed out of place caught his eye. He bent over to get a closer look. "Trees don't usually bleed, do they?"

"You found blood? For real?"

"A couple of red spots on the bark. Could be from an animal or bird."

Drayco looked behind the fence. A row of more bushes flanked it on the other side, adding a layer of potential cover. Even more compelling was what he'd seen on the aerial surveys and was now verified—the area beyond Harry's property had extra privacy from undeveloped parkland on the opposite side of the road.

He took more photos and then hopped back down into Harry's yard beside Nelia and Benny. "The neighbor-witness stated he saw the victim's car pull in around seven that night?"

"Yep."

"The killer would have had approaching darkness helping out, too." What would have happened if Darcie were there along with Harry? Drayco's heart beat faster at the thought of Darcie in jeopardy. But he reminded himself that she was safe and sound, and his relationship with her was over.

They made their way back inside, and Drayco grabbed the book on criminal codes he'd laid on a table. Nelia was right—at first blush, it seemed like an incredible coincidence the victim just happened to grab that particular book. And why was that one page torn?

Even though Drayco had taken photos with his cellphone of the house and yard all along, he made sure to capture the bookshelves and stains one more time. He was used to working with various law enforcement agencies as a crime consultant, but he hadn't dealt with the Fairfax PD lately. They might be collegial, or they might be hostile, you never knew. Best to gather everything he could for his own files.

And yet, it always felt completely natural working side-by-side with Nelia. One thing he was really going to miss was partnering with her in her duties as Sheriff Sailor's CSI. Sure, he'd probably still see her occasionally in some law firm after she got her J.D. in a little over three years. But it wouldn't be the same.

With a last look at Harry's "palace," they left the way they'd come. Drayco made a note to ask Darcie about the symbolism of a statue of Zeus—a mythological figure known for grudges and betrayal—standing in the middle of the giant fountain in front.

Betrayal was an odd theme for a mere landscaping ornament. Was that simply accidental. . .or a conscious choice on Harry's part? And if so, betrayal *of* Harry or *by* Harry?

Benny had scheduled a meeting with his client at the jail tomorrow and asked Drayco to tag along. He wasn't entirely looking forward to it. But then he remembered Darcie's teary eyes, which steeled his resolve. He needed some answers, and Harry Dickerman damn well better have some good ones.

4

Thursday, September 17

After spending yesterday evening poring over the police files and research he'd conducted on Harry, Drayco had rewarded himself with an hour at the piano before bed. But maybe that was a mistake because he awoke later than planned when he didn't hear his alarm go off.

That left him scurrying to get ready since this was one time he couldn't keep Benny Baskin waiting for Drayco's Chariot Service. Their appointment time was not negotiable. Plus, it seemed that Benny had a detour in mind first that would add time to their trip—a diner run.

When they finally pulled in front of their ultimate destination and parked, Benny scooted out of Drayco's car and started to head toward the gleaming modern concrete and glass building. "Thanks for breakfast, by the way. Don't tell my wife I got double helpings of hash browns. She'll want to put me back on that cardboard diet." Benny shuddered.

"I don't know, cardboard has fiber."

Benny just glared at him.

Drayco had brought the Starfire this time and made sure it was locked before he followed after the attorney. It would be his luck to have his car stolen from the parking lot of a jail. Sadly for Benny, there wasn't any time for the attorney's "almost-legal-speed" spin along I-66.

After checking in at the professional visiting area on the second floor of the Fairfax Adult Detention Center, a guard escorted the two men to a small room with a small blue table and three small blue chairs—Benny-sized. Drayco had to hand it to the FPD, though. The place was so immaculate, you could eat prison mystery meat off those floors.

They didn't have to wait long until the guard led a man into the room and shut the door behind him. Drayco recognized Harry Dickerman from the one time he'd seen the guy at Darcie's home in Cape Unity. His gray hair was a little grayer and disheveled, but he managed to be distinguished looking even in an orange jumpsuit, like he should be the host of a TV game show.

Harry sat in the chair closest to Benny. "I saw you three days ago, wasn't it? Do you have a break in the case?"

"Not yet, no. I want to hire Scott Drayco here to assist, but he needs to talk to you in person first."

Harry gave Drayco a close scrutiny, his eyes narrowed into slits. "I'm not sure that's a good idea."

Benny asked, "Why not?"

"I don't need a P.I. because I'm innocent. And if I did need one, this particular P.I. isn't exactly unbiased."

Drayco could feel Benny's invisible side-eye, although the attorney kept his face blank. Drayco half-expected Benny to bring up the topic of Darcie, but apparently, he'd decided it best to avoid the subject and got down to business. "Tell Drayco here how you met the victim, Lara Davidenko."

Harry gave Drayco another piercing stare but complied with Baskin's request. "I met Lara fifteen years ago. This attorney fellow, Stuart Wissler, introduced us. Lara was from Russia, she said. She needed a green card to avoid being sent back to her home country, where she faced sexual abuse. I did the honorable thing and married her at the Justice of the Peace."

Drayco said, "But it didn't last?"

"Six weeks later, she stole some money I kept in a safe. Then the money and Lara disappeared, along with that shyster attorney."

"You say it was the 'honorable thing.' But surely there were other ways to help her apart from marriage?"

Harry licked his lips. "I felt sorry for her. And she reminded me of this other woman. . ."

Drayco waited patiently as the other man looked down at his hands that were trembling on the table. Finally, Drayco prompted him, "Other woman?"

Harry nodded. "A woman I met at a party once. Really hit it off. Another Russian gal. It's not relevant. Too long ago."

"I see. But you hadn't heard from Lara since she and Wissler vanished?"

"Not a peep. I got the marriage annulled through Wissler. At least, I thought I had before he also vanished. But when she showed up at my house five days ago, she told me we were still married."

"Is that why she decided to contact you?"

"I don't think so. The first thing she said when she rang my doorbell was that she needed my help."

"She didn't say what kind of help?"

"Lara seemed shook up, so I went to the kitchen to make her a drink. Thought it might soothe her nerves, you see. But when I got back to the living room. . ." Harry passed a hand over his face. "Sure wish I could get that image out of my head. Lying there. All that blood."

"You didn't hear anything? A scream?"

"Nothing." He pointed at his hearing aids. "Guess I might need a tune-up for these."

Drayco didn't dare look over at Benny when Drayco's earlier guess about their client's hearing turned out to be right. "You called the police right away?"

"I did have the presence of mind to look around the house first to see if the fellow was still there. But I was on the phone with the cops the whole time."

Drayco filed away that timeline for later reference. "During the six weeks you were married, did you not learn anything about her? Her family, her friends, her past?"

"She was always cagey. Guess that should have been a giant red flag. She mostly wanted to go shopping. At expensive stores, of course."

"After she and the attorney vanished, did you try to track either of them down?"

"Yes, but it didn't do any good whatsoever."

Drayco tried not to let his exasperation show, but he was shocked that a man trusted enough to be president of a major international corporation could be so naïve. "You head up Mediasio, is that correct?"

"For the past fifteen years. During which time we've doubled our business." That brought a temporary smile to the man's face.

"Is there a possibility someone associated with a competitor would resort to something as dramatic as this—to ruin your reputation and sabotage your company?"

Harry flinched. "I hadn't thought of that. But. . .I really can't see the world of media advertising being that cutthroat."

"Just something else to consider." Drayco looked over at Benny. "Do you have any photos of the victim, Benny?"

"No and getting one from the FPD could be tricky."

Harry spoke up, "Lara didn't like to have her picture taken. She mumbled something about curses and bad luck. However, there is one photo someone else took, a friend of mine, as Lara and I drove off to the Justice of the Peace."

"You still have it?"

"In the house in McLean. I can tell Darcie where to find it, and she can give it to you." Harry gave one more frosty look in Drayco's direction. "And by that, I mean give it to *you*, Mr. Baskin."

Benny said soothingly, "That would help a great deal. We're lucky you kept it."

Drayco asked, "Did Darcie know about Lara?"

Harry winced, and he picked at the edge of his sleeve. "I, uh, no. I didn't tell her."

"She didn't find out via some other means?"

"Not that I'm aware of. I should have said something, but I didn't think it mattered. I wasn't in love with Lara."

"And you had no idea Lara was making you the beneficiary in her will? Despite having deserted you fifteen years ago?"

He shook his head so fast, there was an audible click. He replied forcefully, "The police asked me that, too. And I'll tell you what I told them. It makes zero sense to me. None of this does."

The guard entered the room, tapping on his watch. Harry gave one last pleading look at the other two men and eased himself out of his chair as the guard prepared to escort Harry back to his cell. Dickerman suddenly looked and moved like a man ten years older.

After the guard had left with his prisoner, Drayco said to Benny, "I've done research on this guy. He seems like a paragon of virtue now, but his father was arrested for embezzling when Harry was younger."

"Probably doesn't mean anything. Good kids come from bad seed and vice versa. But I tell you one thing, I'm going to do some research on my own. About that so-called attorney, Stuart Wissler." Benny joined Drayco in standing up, adding in a big stretch.

They made their way to the front lobby where Benny asked to chat with the lead detective on Harry's case, Shephard King. The man who strolled in wasn't at all what Drayco was expecting—slender build, mostly shaved head, and Drayco spied a hint of a tattooed bicep peeking from underneath the guy's short sleeves.

King shook hands with Benny and then said, "Who's this?"

"Crime consultant, Scott Drayco. He works with me on cases."

Detective King sized up Drayco. "You won't need him. The case is pretty cut-and-dried."

Drayco asked, "You're not considering other suspects?"

"The fiancée, maybe. Even though she has a salon full of witnesses. Could have hired someone."

So, Darcie's fears that she might be a suspect were justified. "What about the book near the victim's body?"

"Book?" King scratched his head. "Oh, I remember now. Looks like she grabbed onto the shelf when she was stabbed. We dusted it." He shrugged. "Checked everything out thoroughly. Standard procedure."

"Did you check the tree in the back of Harry Dickerman's yard? The one in back of the fence?"

"Tree? Why would we?"

"You might want to test it for blood. I thought I saw some specks on the trunk above that fence."

"We didn't see anything to indicate a perp other than Dickerman. No signs of forced entry in the yard or house. Look, any blood you noticed could be a squirrel or chippie. Or a mockingbird or cat. Besides, it rained last night, so whatever you thought you saw likely isn't there anymore."

King stared at Drayco for a moment. "You the same Drayco who solved the Gilbow case?"

Drayco nodded.

"Huh," King grunted. "Pretty clever, with that musical code and all. But I know the cops up in the MPD. They would have worked it out, eventually. And I assure you, we don't need any hocus pocus here."

With that, King essentially dismissed Drayco, pointing his words to Benny before he disappeared through the secure doors, "Better be working on some insanity defense, counselor. Probably your best shot."

Benny practically pushed Drayco out of the building and back to the Starfire, which thankfully was still intact. "Insanity defense, my ass," Benny grumbled.

Drayco opened his mouth to reply, but Benny cut him off. "And don't you start in, either. You aren't convinced Harry didn't do it, are you?"

"With your perfect defense record, Benny, you wouldn't have taken him on if you didn't believe he's innocent. That means a lot to me."

"You're just angling for some of my wife's one hundred-proof fruitcake for Christmas." Benny looked around, and not seeing Detective King said, "King's a hardass, but I found out his wife was murdered ten years ago. Shot during a home invasion burglary. Thought you should know."

"Thanks. I'll keep that in mind."

Seeming to want to lighten the mood, Benny added, "Go earn that retainer of yours. Find out everything you can about this woman, Lara Davidenko. And give it to me yesterday."

It was a sweltering day for the second week of September, and as Drayco cranked up the AC, he had the thought he should go for a dip in Dickerman's ample pool. Hell, even that oversized fountain in front seemed pretty inviting. If Drayco ever did make Dickerman kind of money, he doubted he'd have the fountain or the palace. But the pool? Oh, yeah.

He asked Benny, "What's next?"

"Next? After you drop me off, you go do your magical thing. That hocus pocus. Me, I've got an appointment with a former client I loathe. He was innocent, glad I got him off, but strange bedfellows and all." Benny paused. "I mean, you are on board with this case, right? You did take the retainer."

"Count me in. I may regret it, but my magic will get weak if I don't practice the dark arts."

"That's the spirit." Benny checked the clock on the car's dashboard. "I've got time for a cup of coffee from Hava Java, if you're interested."

"And their S'mores doughnuts?" Dessert after breakfast, now there was an idea. Drayco had seen boxes of the things on the attorney's desk a few times. Make that *quite* a few times.

Benny replied, "Me? Doughnuts? Never touch the stuff."

"I can hear Mrs. Baskin pulling out the cardboard appetizer now."

As Benny opened the Starfire's passenger door, he replied, "You tell her about this, and I'll tell your father you cheat at poker."

"But I don't cheat."

"He doesn't know that. So, what's your first magic act going to be?"

"Maybe not that magical, but I want to learn everything I can about Harry. Due diligence and all."

"Keep me in the loop. And if you find out anything bad—"

"You can dump him so your perfect defense record will still stand."

"I never do that."

"There's a first time for everything." Maybe Drayco had convinced Benny he was joking, but from the way the attorney was giving him a steely glare, he didn't think so.

Okay, so even though Drayco wasn't yet entirely sure of Harry's innocence, he'd promised Benny to try. But was he just hesitant because of his past relationship with Darcie? He'd had cases with stronger personal ties before—even his own mother—and managed to stay objective. This shouldn't be any different, just another day, another case. But god help him, he just couldn't get Darcie's teary face out of his mind.

The universe seemed to be having a bit of fun at Drayco's expense. Maybe not a composer-like-Beethoven-losing-your-hearing level of "fun." But like Beethoven, Drayco's close personal relationships had a way of coming back to haunt him. Would Beethoven have given up composing in exchange for having his hearing back and the chance at a normal family life? Sometimes the fire that burns within consumes all. . .relationships, life, and sanity.

5

After dropping Benny Baskin off at his office—stuffed with coffee and doughnuts—Drayco doubled back along the Dulles Toll Road toward Tyson's Corner. He followed his GPS, but it really wasn't necessary when a tall, modernistic building that seemed to be built entirely out of glass loomed into view. A sign on the side spelled out in large letters, "Mediasio."

He'd done some digging into the company. They owned a couple hundred radio stations in the U.S. and operated tens of thousands of display advertising structures in North America and across Asia, Australia, and Europe. Revenues were measured in billions, which helped explain Harry's palatial home.

Darcie was definitely trading up—unless it all came to naught. Too bad for her they hadn't held the wedding before all of this. But ultimately, it didn't change much. Darcie was who she was, and she seemed to crave a certain lifestyle, the kind Drayco could never have given her. Even if she'd chosen him instead.

After he parked the Starfire, he immediately noticed the strategically placed security cameras monitoring him and any other visitors. The security protocols followed him into the building where two frowning, uniformed guards flanked the front desk, next to elevators requiring access cards. A sign read, "No cellphone photography." Maybe his thoughts about corporate espionage weren't too far off the mark.

Those thoughts were verified when the rep who came to escort him upstairs, Diandre Hoskins, replied to his queries that the security was "Due to all the corporate spying. Though I'm more worried about cybersecurity than physical security, something we're working hard on. As are all businesses. Agile methodology is critical."

On the ride up in the elevator, he asked her, "Regarding that whole espionage angle. Is there anyone who'd want Harry Dickerman gone from the company? A competitor, perhaps?"

"You've been watching too much TV if you think that. Our success doesn't depend upon any one individual. We pride ourselves on our team approach. Synergy and solutioneering is our motto."

Her defensiveness showed through her clenched fists and tightened lips. That was interesting—was she upset Drayco might be impugning Harry's integrity, the company's, or hers? He said, "Then I'm a bit fuzzy on why there would be corporate espionage in the radio and outdoor advertising business."

"You'd be surprised. It's largely China and Russia, but also all the corporate mergers. Bigger companies look for ways to swallow up smaller companies. Any weaknesses would be fodder for that. Plus, our firm is innovating all the time. There might be others wanting to scoop some of our patents and ideas."

"What sorts of patents and ideas?"

She cocked an eyebrow. "Now *you* sound like a spy. Products, marketing, you name it. Bleeding edge stuff. And we're not eager to open *that* kimono."

The elevator stopped on the twelfth of twelve floors. It opened to reveal a walled-off secretary's station with a door at one end. Hoskins waved at the woman seated there and led Drayco through the end-door. The space behind revealed an entire floor dedicated to the CEO.

The main area had wooden paneling in the middle with a Scandinavian-style white desk in front and a sleek black leather sofa backed up against one wall of windows. The space's opposite side had a glass-enclosed meeting room with a white marble-topped table and twelve chairs on top of a mini-platform, also white.

Diandre Hoskins handed him off to the acting CEO, Sloan DelRossi, who wasn't wearing all-white, thankfully. But his perfectly coiffed silver hair and designer duds and glasses made him look like he'd popped off the cover of *Fortune.*

DelRossi had barely shaken Drayco's hand when the first words out of his mouth were, "You think it's me or one of our employees, don't you? My wife loves all those police and CSI shows. I know the drill."

"We're just looking into Harry's background to find whatever we can to prove his innocence."

"If it helps, I handed in my retirement notice a month before this whole nasty business. I may have to stay on a little longer, alas. But my wife has plans for me to set up her long-awaited koi pond. So I don't think I can delay much more." He laughed.

"Who would be next in line for CEO, then?"

"We have a few possibilities but no frontrunner. And there's some preliminary talk amongst the board to merge with another company. It's all in a state of flux. But time's critical, and we can't afford to boil the ocean too long."

"I chatted briefly with Ms. Hoskins about corporate espionage."

DelRossi waved a hand in the air. "There are always bad actors, but I doubt that had anything to do with Harry's situation. To be honest, I suspect it's a domestic thing. Those are more likely to take down CEOs. Whether it's cheating or sexual harassment or whatever."

"Do you have much of that here?"

"There have never been any hints of improper behavior with Harry or complaints against him. Harry is well-liked by everyone, even his competitors. He's a real strategic thinker."

"Did Harry ever mention the victim, his ex-wife?"

"Didn't know anything about her, so that's a big no. I firmly believe she was a plant. Or maybe she was there to murder Harry for some reason, burglary perhaps. And her accomplice did it."

"You're that sure of Harry's innocence?"

"I'd bet my life on it."

Ordinarily, if someone spouted a clichéd response to Drayco like that, he'd be skeptical—since it was usually the opposite sentiment in private. But in this case, DelRossi's conviction came across as sincere. That made three people so far who were convinced Harry wasn't guilty.

DelRossi went to a window to close a blind. Symbolic, perhaps? "Besides, Harry's grasp of the company's core competency and his command of our performance management system is unparalleled. Ask anyone."

Drayco tried not to show his exasperation at all the corporate-speak. Having heard it roll out of Diandre Hoskins's mouth like anti-poetry was bad enough. "Harry was based at this building, is that correct?"

"Yes, although we have satellite offices in Maryland, the District, and far beyond. But he spent most of his time here."

When DelRossi got a call he indicated was important, Drayco let himself out through the door and back into the secretary's area. As he waited for Hoskins to come up and escort him down again, he noted the nameplate, Cherry Wesby.

He said, "Did you work for Harry Dickerman, too, Ms. Wesby?"

"Yes, I've been here as long as Harry has."

"Must be difficult with his arrest."

Her eyes teared up. "I'm beyond devastated. I enjoyed working for Harry so much. He always remembered my birthday and my husband's and kids' birthdays, too."

"Do the other staff feel the same way?"

"Absolutely. Harry is such a generous man. He once helped out an employee who lost everything in a fire and wasn't adequately insured. Even paid his extra medical bills."

He thanked her as Hoskins arrived to take him back down to the lobby. Upon exiting the building, Drayco was conflicted. He was getting the impression Harry was a paragon of virtue. Saint Harry, indeed. Was the guy an amazing actor leading an amazingly secret double life, or was he truly innocent? Harry's secretary had mentioned "other staff" who were also devastated.

Drayco felt increasingly sucked toward the innocent camp himself, but he wasn't ready to go all in. The mysteries surrounding the victim were far too anomalous for that. It was just too convenient she was killed right after Harry got engaged to another woman. Or was it?

Once outside again, Drayco went around the perimeter of the building, taking photos. That would give the security staff something to talk about. On his way back to his car, he noted a tall billboard on the other side of the street touting a popular insurance company. He could just make out the smaller print at the bottom of the sign, "Mediasio Advertising."

What would Hoskins or DelRossi say? Drayco mouthed aloud, "That sign is a fine expression of solutioneering synergy. With a dose of frictionless paradigm-shift empowerment, if I ever saw one." If he had a job where he had to speak business-babble all day, he doubted all the survival strategies in the world would save him. It would be a fate worse than death.

He hadn't got what he'd come for, a good lead on a possible business rivalry or competitor who hated Harry enough to want him out of the picture. Not that he was loving that angle, anyway, since how would a "ghost" fake ex-wife figure into it all? Then again, she was Russian, wasn't she? Diandre Hoskins had mentioned China and Russia as objects of their cybersecurity defenses.

Drayco knew a couple of things, at least. One, he was hungry and wished he'd snagged some doughnuts for himself. And two. . .he wasn't entirely sure he *wanted* Harry Dickerman to be innocent. Disgusted with both the hunger and the whiff of pettiness that surprised and irritated him, he scurried off to his car. Doughnuts first, and an apology to his professional code of conduct later

6

Friday, September 18

The day dawned sunny and bright with a hint of flamingo-colored clouds and a light wind. Perfect for flying. The short flight from Virginia to Maryland had taken them over the Chesapeake Bay with views of the small islands and their narrow, caramel-colored beaches below. But those picturesque views were marred by the reddish-brown "mahogany tide" caused by algae blooms starving the water of oxygen. Farm fertilizers filled with phosphorus and nitrogen made for a deadly marine mix.

After they landed at their destination, Drayco buttoned up the Cessna Skyhawk and headed over to the Salisbury airport terminal to pick up his rental car as his companion tailed him, chatting all the way. "How's the Opera House project going?"

"Don't ask."

"That good, eh?"

Drayco replied, "Just swell," hoping that would get Mark "Sarg" Sargosian off his case, but the man piled it on. "Okay, so how's the concert prep going? Aren't you supposed to play something for that scholarship thing at UMD?"

"I don't want to talk about that, either."

"If I'd known I was flying Air Grumpy today, I'd have reconsidered my offer to tag along."

"Sorry." Drayco added in a more conciliatory tone, "The Opera House renovation project is stalled since I've run out of funds. Companies are more tight-fisted with arts funding these days. It's like a professional wrestling match when it comes to groups competing for the few grants." That made him think of Nelia's reply that there weren't enough grants to go around for her, either.

Sarg said, "Ouch. Sorry I asked."

"And about the whole UMD thing. . .it's only because my uncle set up that scholarship in my name they're half-interested in me performing. As if I could ever be ready for a recital."

"Why not?"

"The right arm. Too unpredictable. And I'm not going to play pieces for the left hand only. What's the point?"

Sarg shot him a sympathetic look and finally let the matter drop. He turned the conversation back to the reason they were here, the murder victim's neighbor, Jilliana Vaughan. "Think she'll talk to us?"

"She said she would. Guess we'll find out."

"Thanks for letting me come along, by the way. It's a welcome change."

"Paper cuts and phone-ear-burn not doing it for you?"

"Everybody thinks the BAU is like on TV. Rushing around in our private jets to play cop and nabbing the unsub. All in sixty minutes."

"Gets the attention of the ladies when you tell them who you work for, though."

Sarg grinned. "Tell that to my wife. Having an FBI-agent as a hubby doesn't impress her at all. Now, if I really want to impress her, I'll fix that mantel clock that's been busted for ten years."

Drayco followed the directions Jilliana Vaughan gave him over the phone and in short order, they drove up to a series of four-story buildings set on the banks of the Wicomico River. They were reasonably well-maintained units, albeit on the bland side. With a focus on comfort, not controversy, they were the architectural equivalent of khaki slacks.

The apartments dated from the late '90s when developers bought up the land, and high-density housing began popping up along the river like noxious hogweed—bringing nitrogen and phosphorus with it. More "mahogany tide."

They knocked on the door of unit 306, and Drayco introduced himself and anointed Sarg as his "associate." Jilliana Vaughan led them inside to a bright, airy space transformed into an art gallery, or so it appeared. A montage of carefully placed prints in various sizes set in colorful frames perched above a low-slung gray couch with a daybed. Even the flooring consisted of one geometric rug in shades of gray at one end and a Native American Modern Crystal mat in oranges and reds at the other.

Jilliana herself seemed coordinated with the room, with beads in orange, red, and silver gracing the ends of her cornrows. She seated them in two chairs with abacus-like seats and headed toward the sofa but first stopped to straighten some books on a side table, so they were in perfect alignment.

She sat down as if there were tacks beneath her. "I have to tell you this whole thing is like a bad trip."

"Did the police interview you?"

"Some dudes from Virginia. Fairfax, I think." She reached up to fiddle with the beads. "Chatted with them only as much as I had to. Didn't want to get involved. Certainly not with the cops. I had a family member. . ." She hastened to say, "I mean, I'm sorry for Lara and all. She didn't deserve that."

"We're not the police. What you say to us is more likely to be confidential."

Her shoulders relaxed, and she stopped fiddling with her beads. "Don't know what I can tell you. She hadn't been here that long."

"How long?"

"Three weeks."

"Did you chat with her much?"

"We went out for drinks. I was surprised her accent wasn't all that strong, her being Russian and all. She didn't say a lot about herself. Oh, there was this one thing, after she'd had a few beers. I didn't tell the cops this. She asked me if I'd heard of Synanon. And then she said she was involved with something worse."

Drayco gave a quick glance at Sarg. "Synanon as in a cult? Did she give the name of this group?"

"Yeah, but I didn't write it down or anything. Started with a 'G.' G-something Farm, I guess." Jilliana repositioned some pillows beside her until they were perfectly flush against the sofa. Drayco suspected OCD rather than nerves.

He asked, "Did she add any more details?"

"Nah, but I guessed it was some of that leftover hippy thing. To be honest, Lara was kind of jumpy, easily startled. Kept her shades drawn. Had to kidnap her to get her to go out for drinks."

"She didn't say what was bothering her?"

"Just that she was in some trouble. Didn't have anyone to turn to. Except the only man who'd ever shown her any kindness."

"When did she mention this?"

"The day before she was killed. Called me on my phone. Said she was going to see him the next evening around six or sevenish. When she thought this guy would be home. More like after she worked up some courage, I'd say. I offered to drive her since she seemed upset. But she said she'd be fine and drove herself."

"Did she mention the man's name specifically?"

"Harry Dickerman."

"Not some other Harry or another man?"

"I'm pretty sure she said it was the guy, Harry Dickerman. The one who killed her, or so the cops say, right?"

"He's the one arrested for it, yes."

Jilliana chewed on her lip. "You know, it's funny. That same night she got a bit sloshed, I got the impression she wanted to move to Russia or some other country."

"Did she say why?"

"Get a clean start. But she wasn't sure she'd be safe there. Or anywhere, for that matter. Said she'd been saving some money." Jilliana arranged the pillows again and gave them a little pat. "Guess I felt bad for her. She also said she'd made mistakes and wish she had a do-over from her childhood. Sad, huh?"

Drayco thanked Jilliana for her time and asked if the building superintendent might be available. She replied, "Karl Hondo's his name. Manages the place. You can try him, he lives in 121. Next door to the laundry."

As Drayco and Sarg took the stairs down to the first floor, Sarg looked over at him with a slight frown. "You have piles of sandbags under those icy blue eyes of yours. Got some of that hypnagogic sleep paralysis thing going on again?"

"Not exactly. Lots of bad dreams, though. I keep falling into a deep pit of water and nearly drowning."

"The same dream?"

"Variations thereof. I've had them since this summer after that suicide-sonata case."

"Likely stress. Dreams don't mean much other than stuff that happens thrown into a giant brain blender."

"That's not what the Akashic Records would say."

"The woo-woo thing the kids on that suicide case believed in?"

"Loosely believed is more like it. Tapping into the universal encyclopedia. Able to see the past, predict the future."

"Falls under woo-woo to me. But those dreams, man." Sarg studied his face. "You mentioned falling into a deep pit. You still have touches of claustrophobia?"

Drayco stepped down hard onto the first-floor landing and almost tripped, saving himself from a bruised ego by grabbing the railing. "Not for a while. You're the only other person who knows about that, by the way."

"Guess we all have stories of bratty kids when we're young. Okay, maybe not as nasty as your cousins who locked you inside a footlocker for hours."

Drayco took a deep breath. *Keep that in your mental lock box, Scott, along with all the other horrors of your life.* "They apologized years later. And it's ancient history."

"Makes it all the more impressive you can get into a small plane and fly it all over the place."

"You'd be surprised how many pilots have claustrophobia. And acrophobia, too."

"A fear of heights? Huh. Pilots either have a secret cure, or they're insane."

"A little of both."

"Well, everybody has some fear of one kind or another."

"Yours being peanut butter and mayonnaise sandwiches, as I recall."

Drayco was rewarded by the look of disgust on the other man's face. "Every time you had one of those, I wanted to barf. You ever do that again, and I'll make sure I barf on you."

"Just don't barf in the plane."

Sarg grinned. "Thanks for offering to drag me along today. You don't really need me."

"Pshaw."

"Pshaw? Who still says pshaw?"

"My grandmother used that word all the time. I kind of like it. Besides, it's a lot nicer than codswallop."

That got a belly laugh out of Sarg, with all of its O-shaped indigos and teals with cottony edges. Maybe that's why Drayco liked the guy so much—there weren't many sounds with the same synesthesia combination of colors, shapes, and textures in his voice. If Nelia's voice was audio honey, Sarg's was a therapeutic salve for Drayco's brain. And he needed all the therapeutic salve he could get in his line of work.

Sarg salve. . .kinda sounded like a tonic for a chronic case of nerves. Too bad it couldn't soothe a rash of lies and deception. Or murder.

7

After their conservation with Jilliana, Sarg had teased Drayco about referring to Sarg as his "assistant." But when it came to the building's manager, Drayco was more than willing to have Sarg pull out the official card. When Sarg flashed his FBI badge to Karl Hondo, the man was quick to take them back up to the victim's apartment, this time via the elevator.

As he put the key in the door, he said, "Be my guest, but the police already checked out Lara's flat. And by that, I mean both the cops here in Salisbury and the guys from Virginia. Didn't find nothing much that'd help, near as I could tell."

"Any hints of what they did find?"

"To me? You kidding? Besides, all I cared about is when they told me they was done." Hondo thrust the key in the lock but turned back to Drayco and Sarg. "What they didn't tell me is about any family coming. You know, to pick up her things. Guess I'll have to have a yard sale or something."

The man laughed, then quickly grew more pensive. "Seriously, though, if you guys can find a way to clear all of this out, be my guest. She still has five weeks on her lease. Paid in advance. So I'm not hurting. Not yet."

Drayco asked, "No references required to rent here?"

"Nope. We ain't The Ritz."

"Did she say anything to you when she signed the lease? About any friends, family, or her background?"

"Didn't say nothing."

"Anything about her strike you as odd?"

"The only thing I cared to know was I got a nice envelope full of cash in my mailbox for the rent."

"Cash?"

"Yep. Hey, I gotta make a living, you know? As long as the rent keeps coming, I pretty much leave 'em alone. Don't like police coming 'round, though. If I have a feeling cops'll wind up here, I don't rent out. But this woman, she seemed normal, so I said okay. Plus cash. With checks, you always gotta worry about 'em bouncing, you know?"

He opened the door and immediately exclaimed, "*¡Ay, Dios mío!*"

The three men stepped inside the apartment that had quite obviously been trashed. Papers cluttered the floor, chair cushions were tossed around, cabinet doors were open, and overturned boxes spilled their contents.

Drayco said, "Guess it wasn't like this after the cops left?"

"Hell, no. Would have had their badges for it."

Sarg piped up, "Seen anyone suspicious around recently?"

"We get creeps from time to time, sure. Haven't seen any past coupla days." Hondo shook his head. "I'll check with her neighbor next door. See if she heard anything. And call the cops."

Drayco decided "divide and conquer" would be the best course of action, so he let Hondo check with Jilliana while Sarg and Drayco surveyed what was left of the victim's apartment. If it weren't such a mess, it would look roughly the same as Jilliana's, though with a spartan decor in black, white, and beige.

He looked over at Sarg. "I'll have to alert the Fairfax PD, too."

"Figures." Sarg paused and then added, "Bring any gloves with you?"

Drayco shook his head. "I keep them in the car, as you know. Didn't think I'd need them in the plane."

"We can look from afar but not touch, then."

They stood close to the doorway, trying to see as much as they could from that vantage point. Finally, Drayco said, "Act first, apologize later," and he slipped off his shoes and headed inside the living room, careful not to step on anything.

There wasn't much in the way of personal belongings. A paltry few clothes, some toiletry items all smelling of lilac, plain brown dishes, a deck of cards, and a multi-colored, hand-painted wooden egg with a starburst design. He spied a few books and checked the titles to see if any matched the one found next to the victim's body at Harry's house, but no luck. Just a few romance novels and a book on how to write a bestseller. Did she pen any fiction, maybe based on her own life? Or better yet, an actual memoir?

He did find a small wallet on a table that he flipped through with his multi-tool to avoid leaving prints. It had exactly one photo, of a younger Harry. Sarg came over to look. "That who I think it is?"

"Our client. Seems to be the same vintage as the photo of Harry and the victim—the photo Darcie got for me from Harry's palace."

"The victim kept a copy all these years? Odd for a 'sham' marriage, don't ya think?"

"Maybe when Lara told Jilliana she was turning to the only person she felt she could trust, namely Harry, she meant it."

Sarg tugged on his earlobe. "You sure you can trust Darcie? She could have killed this ex-wife when she found out about her. Then Darcie trashed her apartment out of jealousy."

"She's got an airtight alibi."

"A convenient one. But only for the murder." Sarg waved his hands around at the mess. "This is a whole 'nother thing."

Drayco grabbed his cellphone to call the Fairfax police. After giving them the few details he knew, he hung up to do some quick internet research on his phone. Sarg asked, "Whatcha checking?"

"Found something interesting."

"What's interesting?"

"Gaufrid Farm. A commune on the Eastern Shore, half in Maryland, half in Virginia. Matches the partial name Jilliana gave us."

"What kind of commune? Like the Synanon group she mentioned?"

"I hope not. You remember them, right?"

"Sure. Lovely bunch of folks. Drug rehab turned violent cult. Attempted murder, legal problems, terrorism. Fun."

"Gaufrid Farm hasn't been on the Bureau's radar for any of that, has it?"

"Never heard of it." Sarg pointed to Drayco's phone. "What does that say?"

"It's a commune for former Wall Street employees. Love and peace and brotherhood and all that."

"What would a Russian woman, who conned someone into a sham marriage on the premise of sex trafficking, be doing in a commune for Wall Street types?"

Drayco frowned. "I don't know. Yet."

Sarg sighed and checked his watch. "Wonder how long it'll take the cops to make it?"

"Sorry about that. Didn't know your luncheon trip was going to turn out to be an all-day affair, did you?"

"I *adore* cross-jurisdictional cases. This is turning into a megillah."

"You're telling me."

"And here I thought I was just coming along for good seafood."

Drayco raised an eyebrow. "Just the seafood?"

"Okay, and the whole Russian angle may make it a Bureau case soon. If trafficking is involved."

"Really?"

"How do you think I got Director Onweller into letting me come today?"

"I wondered about that."

"Hell, I think he still secretly hopes you'll return to the Bureau someday."

"He might be waiting a long time." Drayco kicked over a sofa cushion on the floor with his foot. "Hate to agree with you, but the Russian connection does add weight to the whole human trafficking thing."

"Or money laundering."

"Or mob."

"Yay." Sarg's eyes had a new gleam in them, and Drayco almost felt sorry for his former partner. The agent's cases must be a tad on the routine side lately.

As he continued to look around the room, Drayco noted the curious absence of any tech. No cellphone, no computer, or tablet. Maybe that's why the glint of something shiny caught his attention. The shininess emanated from beneath the overhead light fixture, but the fixture itself wasn't silver at all—just black and white. Drayco scooted a nearby chair over with his foot until it was underneath the light.

Sarg asked, "See something?"

"Perhaps."

Drayco stood on the chair and got as close to the source of the shiny reflection as he could without touching it. "Well, well."

"What is it?"

"Why don't you verify what I think I'm seeing."

Sarg dutifully switched places and lifted up on his tiptoes to crane his neck and get a good angle. "Looks like a bug."

He hopped down, and Drayco replied, "Yep, that's what I thought."

"Wonder if our mystery room-trasher put it there?"

"Before or after the trashing?"

"Either."

Drayco got up on the chair again long enough to take a cellphone photo, then hopped back down. "After doesn't make much sense. Before, on the other hand, well. . .it's possible. But if they did, and it was the same perps as the trashers, why didn't they retrieve it when they came back to search the place?"

"Someone or something interrupted their ransacking game?"

"If so, perhaps they were planning on returning tonight. That is, if they didn't find whatever it was they were looking for."

"Too bad for them. But why would someone bug Lara in the first place?"

Drayco took a few more photos of the messy room. "Unknown. It does mean one thing. . .whoever was listening in overheard Lara's call to her neighbor the day she was killed."

"That she was going to see Harry, and at what time."

"Quite possibly, and he followed her. Maybe the perp had plans on killing Harry, too. But when Harry went to make the drink in the kitchen, it gave the killer an opportunity to take out his primary target and frame Harry."

Karl Hondo popped back into the room. "Think I saw the cop cars in the parking lot. Should be up here soon. The neighbor says she didn't hear anything weird. But she likes to listen to music with the sound turned up. Turns out she's a Tito Puente fan like me."

The men waited for the arrival of the local police with equal parts anticipation and resignation. It would certainly put their schedules into disarray, as Sarg had mentioned. Still, it would be interesting to see what the police made out of the listening device.

To Drayco, that was the most curious aspect of the ransacked apartment. He doubted the bug was from a previous tenant since it was way too coincidental for that. Jilliana said Lara Davidenko seemed jumpy, almost afraid. That meant someone was very much interested in Lara, enough to stalk her, bug her, and silence her permanently.

Drayco had to admit—grudgingly—it was a stretch to imagine Harry Dickerman going to such lengths when there were simpler ways to handle the problem. Although corporate espionage involved spy tech sometimes, did it not?

He shook his head in wonder. *What the hell had you got yourself into, Lara Davidenko?*

8

Thankfully, the police encounter at Lara's ransacked apartment hadn't taken as long as Drayco and Sarg feared, and they'd managed to get in the seafood feast Sarg was drooling over. It was a very late lunch, but still technically lunch, heavy on the pan-fried calamari with hot cherry peppers and the achiote spiced fish tacos.

After flying back to Manassas and dropping off Sarg at the VRE station to catch a train, Drayco made his way to his small D.C. office for an appointment. He grabbed some papers threatening to fall off the top of the vintage turntable and dumped them into a drawer. He tossed out some takeout containers, pushed some filing boxes under the desk, and wedged a loose pile of books between two treble clef-shaped bookends. He stopped to critique his handiwork. No more Leaning Book Tower of Pisa, anyway.

He hadn't had many visitors to the office in a while, and it showed. Maybe not as bad as Lara's trashed apartment, but not pristine, either. He really should hire a maid service.

He took a sniff around. Should he use some air freshener? No, too obvious and too piney. Instead, he hastily turned on the coffee pot to brew some Kona coffee—refreshments and a caffeinated "incense" all in one.

The woman who'd made the appointment had been insistent. She wouldn't tell him what it was about, just that he'd want to hear her story. When the phone rang, he half-expected it to be her canceling, but it wasn't. "What's up, Benny?"

"Wanted you to know I can't find boo about Stuart Wissler as an attorney. Not in any Bar Association. No bar records whatsoever in the whole U-S of A. And none of the area law schools have a record of him." He added with a wry tone, "Why, I'm beginning to think he wasn't a real attorney at all."

"Sounds as phony as Lara Davidenko's marriage to Harry." Drayco filled Baskin in on the visit to see Lara's neighbor and then Lara's trashed apartment.

"Poor baby. Sounds like your little jaunt turned out to be an ordeal."

"The local police released us pretty quickly. I've already chatted with the charming Fairfax Detective Shephard King, who was not a happy camper."

"I'll bet. What did he say about that listening bug?"

"Not much. Although I got the impression the good detective thinks Darcie is complicit in all of this."

"Really? Darcie using a high-tech listening device? He's *met* Darcie, hasn't he?"

When Drayco heard a knocking on his office door, he rang off with Benny and ushered his visitor inside. Alisa Saber looked to be in her mid-twenties, with a waif-like face and long blond hair fringed with bangs. She was also wearing an earnest look and a big chip on her shoulder.

She plonked down into the purple tufted chair across from his desk and immediately asked, "You're looking into the Harry Dickerman case, aren't you?"

Drayco blinked at her. "That's not common information."

"I saw a news article on the internet. Benny Baskin was listed as Harry's attorney, and I did some research. He uses you on his cases."

"Assuming that I am indeed working on this particular case, what's your interest?"

"I think Harry is my biological father, and I want to hire you to prove it."

Drayco sat up straighter as he took a moment to shake off that bolt out of the blue. "You don't need me for that since a simple blood test will do."

"But he's in jail. So I can't just walk up to him and ask for one. Besides, I want you to find my mother, too." She wiped her palms on her pants as if they were sweating.

Drayco pulled out a notepad. "Why don't you start at the beginning."

"Six months ago, I did one of those online ancestry DNA tests and found out I have two cousins. One is linked to Harry Dickerman via his brother, Mike. Mike Dickerman is deceased. The other is Olga Whitman."

"Did you contact both of these cousins?"

"Yes, and Mike's cousin was open with me but didn't have much to offer because Mike died fifteen years ago. Plus, he'd had a vasectomy after his kids. So I couldn't be his child, but the DNA showed I was clearly related closely somehow. I tried corresponding with Olga via email but only got a vague reply. Something about being brought to the U.S. from Russia by an adoptive family."

Drayco's ears perked up at the Russia reference. Was there a connection to the murder victim, Lara Davidenko, also Russian?

Alisa pushed the bangs out of her eyes, revealing a small, almost heart-shaped mole on the side of her nose. "After I heard about Harry's arrest in the news, I decided to try you first. I don't want to get caught up in the legal part of things yet. If I hire you, you'll keep my secrets safe if I want you to, right? All that client privilege stuff?"

"As much as I am legally able, yes."

She sighed. "Okay, I guess."

"Do you have any other information about either of your biological parents?"

"Other than the ancestry DNA records, not much. Nothing about my mother at all, but I emailed a former worker at the agency where my parents adopted me. I heard them discussing that place once when they didn't know I was listening. I asked my mother later, and she said I was literally dropped off there."

"Adoption agencies aren't allowed to give out personal details."

"Oh, I know. But I found out the former worker is retired. She told me she remembered my case particularly, because the woman who dropped me off spoke with a thick accent. Was hard to understand. After handing me over, she also handed the ex-worker a scrap of paper with a few words written down on it. The worker didn't understand that, either."

"Did she recall what those words were?"

Alisa handed over a printout. "No, but here's a copy of the DNA tests. I attached the ex-adoption worker's name, too."

Drayco grabbed a new manila folder and slipped the papers inside. "You're what, twenty-fourish?"

"Twenty-five."

"I'm curious. Why are you trying to find your biological parents after all these years?"

Alisa's hands always seemed to be in motion, this time as she knotted and unknotted her fingers. "My adoptive parents got divorced ten years ago. My mother remarried and moved to Sweden when I was eighteen. My father died at age forty-seven from a blood clot due to an inherited gene. So I realized I have no idea about my own lineage. What if I've got some hidden genetic thing, too?"

"That seems like a reasonable way of looking at it."

"Plus, I'm a biology grad student, so this made me interested in genetics." She pushed her bangs out of her eyes again, and he wasn't sure she even realized she was doing it.

He replied, "There is a slight possibility this could make you a suspect in Harry's murder case."

"I don't care." She looked away for a moment. "I tried to imagine what my biological parents were like. Harry isn't that far off, except I thought he might be powerful. Like an entertainment guru or sports celebrity."

Drayco waited patiently until Alisa turned her gaze back to him, but she seemed to be struggling to find the right words. He gently prompted her, "And your mother?"

"My natural mother, on the other hand, would be an aspiring actress or a wannabe politician who had me when she was young. Maybe fifteen or sixteen and couldn't handle the responsibility of a baby. I pictured her as beautiful and smart and strong and funny and wise. And I imagined all the birthday presents she was saving up for me in case we ever met."

Alisa paused again, and he looked into her eyes, seeing the first hint of tears. Should he get her a tissue?

She continued in a softer voice, "Most people can't imagine what it's like to never know your mother. To have her missing out of your life."

Flashes of awkward birthday parties when Drayco was a boy washed over him. How he and his father studiously ignored Mother's Day. And how he always looked out at the audience during his youthful piano concerts, hoping he'd see one particular female face, but she never appeared.

He took in a deep breath. "My rates aren't cheap, but seeing as how you're a student, I can cut you a deal."

"You'll take the case?"

"I promise I'll look into it. But I can't promise what I find will be exactly what you wish it to be."

For the first time, a look of uncertainty flitted across her face. "Knowledge is power. I just want to feel like a whole person again." She pushed her bangs out of her eyes again, her next words showing more of her spine. "Besides, as my adoptive mother likes to say, you gotta challenge your fears."

He got up to hand her some tissues before showing her out, then sank into his desk chair to glance absently at the notes he'd jotted down. Alisa had pegged his empathy meter off the charts with her missing-mother story, but he didn't like to take cases due to that. They often turned out badly.

On the other hand, he couldn't deny this had to be more than a coincidence with her turning up now. Was it a shady scheme on her part with Harry in the middle of it all? The fact there was another Russian connection was also intriguing, and he'd be a lousy detective if he didn't pursue that angle.

He started to grab his notepad to take home with him, but instead, he stopped and pulled out his wallet. He slid out a faded thirty-year-old photo and studied it for a moment. The auburn-haired woman stared back at him, a ghost from his long-ago past—and briefly, his recent past. She smiled ruefully—as if she could foresee the crater she'd leave in their lives when she left Drayco and his father behind.

He put the picture back in the wallet and grabbed his notepad. He had two cases to deal with and no time for maudlin trips into places that never were or never would be. Besides, he had a hell of a lot more research to do because tomorrow there was going to be some "farming" in his future.

9

Saturday, September 19

Drayco got lucky with the morning traffic heading east over the Bay Bridge, more of a caterpillar crawl than a full-stop for a change. He took a few wrong turns before he finally wound up in front of his target, marked only by a small wooden sign next to a long gravel driveway that said, "Gaufrid Farm" in faded blue letters.

He'd already studied the place from satellite photos. They showed the seventy-acre property straddling the border between Virginia and Maryland on the Eastern Shore, although the main building complex was in Virginia. The land backed up to a seemingly endless marsh and was a mile away from the Chesapeake Bay, but it was as far from Wall Street, atmosphere-wise, as you can get. Maybe that was the point.

As he pulled into a parking space on the sandy-soil lot, he had to dodge a truck with lettering on the side that said Lavigne Brewery Supply, where a uniformed driver was unloading gear. To the right of the truck sat a small aluminum jon boat, buttoned up and perched on a trailer.

Drayco glimpsed a building with corrugated siding, a green roof, and wooden posts—an office, perhaps?—at the end of the lot and headed for that. When he ducked inside, the only occupant was a man with a head full of silvery-white hair and a neatly trimmed beard to match, who was flipping through files in a cabinet.

He stopped and turned around, saying in a voice that was a little hoarse, "Visiting days are on weekends. We don't allow them any other time."

"I'm Scott Drayco. We spoke on the phone."

"Ah, yes. You described yourself as a private law enforcement consultant, I believe?" The man pumped Drayco's hand with the grip of a man twenty years younger. "I'm the director here. Gordon Aronson, at your service."

"Your property straddles two states, doesn't it?"

"And we get to pay taxes in two states. Although Maryland receives a small fraction since the main buildings are in Virginia. Forgive me for asking, but I find it odd that a private detective is interested in property boundaries. Unless there's a lawsuit coming to challenge our legal status?"

"Not property-related, no."

"That's a relief." Aronson cleared his throat, although his voice was still raspy. Without prompting, he apologized, saying, "I like to say I got hoarse from yelling at people. But I had surgery for a benign esophageal tumor last year."

"I'm glad to hear it was benign."

"Yes, well, as an ex-smoker, could have been worse. But I doubt you came all the way from Washington to chat about my health."

Drayco gave a half-smile. "I want to discuss a former resident of yours, Lara Davidenko."

"Ah, Lara. We were sorry to see her go. Been here for fifteen years."

Drayco took note of that date. Fifteen years was the same time frame as when she bilked Harry and then "vanished." He said, "Do you have any idea why she might be murdered?"

Aronson stared at him with unblinking eyes. "Murdered? I had no idea she was dead, much less murdered. It's only been three weeks since she left."

"You hadn't heard about her death, then? From another member or from the news?"

"I try to stay away from the news, one of the reasons I'm here."

"Has no one contacted you yet? Family? Law enforcement?"

"No, you're the first, Mr. Drayco." Aronson stood swaying in one spot, and Drayco was afraid for a moment the man would topple over.

He reached out a hand to steady him, but Aronson straightened up and seemed to pull himself together.

Drayco said, "I know this comes as a shock. I'm sorry for the loss of your friend and colleague."

"I'd gotten used to the loss of her presence. But this. . .it's different when you know you won't ever have the chance to see someone again."

He picked at his cuticles, seemingly lost in thought. Then he added abruptly, "Time for me to make my rounds. Want to join me for a tour? I need the distraction, and it's either that or hit the bottle. We can talk more as we go."

They strolled toward a garden patch where Drayco got a peek inside a kitchen in a small building as they passed by. Aronson pointed to the garden. "We grow much of our own food here. Not all, of course. In fact, two of our members are currently at the local market."

The side of the kitchen building had a fresh coat of pewter-colored paint. When Drayco made note of it, Aronson explained, "We've had thefts, graffiti, occasional threats."

"Have you reported them?"

"Just thought it was kids getting their kicks. It's not one of our own, I can tell you. We all use the same car, and it's out in the open. Wouldn't be easy for any member to steal anything." He stopped walking for a moment. "It couldn't be related to Lara. Could it?"

"Hard to say." Drayco stopped to examine the new paint job. Very professional. "I'm curious. . .why did Lara choose to live here? I thought this was a commune for ex-Wall Street employees?"

"She wasn't a Wall Street worker, no. But she'd fled from her evil Wall Street master, who made her a sex slave. The fellow's now dead, by the way, or so she said."

"Why didn't anyone contact the FBI about this so that the man could be prosecuted?"

"This is a place where people with secrets and guilt can escape from their pasts. And Lara seemed desperate to get away from her trauma."

"Why did Lara leave the commune right before her death? Far too coincidental, isn't it?"

"She said she needed a change." Aronson resumed walking and motioned for Drayco to follow. "I told her it's much safer here. We counseled her that returning to normal life would be dangerous. Look at all the gun violence."

Okay, so the man kept up with the news somewhat, so why hadn't he heard about the victim's death? Drayco asked, "What is your purpose here? Your mission?"

"Twenty-five years ago, I started this commune for ex-Wall Street refugees like myself. Spent a lot of capital building up the place because I believed in it. Still do. But. . ."

Drayco waited for him to add more, but when he didn't, Drayco prompted, "Yes?"

"Our numbers keep dwindling. Most members have trickled away. A few due to natural means, but others returned to the trade. You think heroin is addictive, try the highs you get from the adrenaline-fueled world of trading."

"You said natural causes. Are you sure of that?"

"We lost Wesley Brewis to a brain aneurysm. Niles Peto of a cocaine overdose, sadly. Boyce Hershorn, ex-commune member, left to start up a law practice again. Eventually ran for office, first as a state representative and then a U.S. rep a couple of years ago. Daven Monk, another ex-commune member who'd been here for ten years, left not too long ago to join a monastery. He worked in our brewery here as an assistant, but I think he's into cheese now."

"A monk whose last name is Monk?"

Aronson chuckled. "I know. We thought that was rather appropriate."

"Lara's departure must have hit you hard, following as it did on the heels of those losses."

The older man kept up a steady pace and seemed hardly out of breath. "Lately, I've had to wonder if it's all been worthwhile, this project of mine. Even before Lara left, our group was on its last legs, to be honest. I'd hoped to recruit some more former Wall Street escapees. After all, the next big crash is coming. I mean, isn't it always?"

"And when it does, you'd be able to convince more people to try out the commune?"

"Better than throwing yourself off a building."

"How many members are left?"

"Max McCaffin, short for Maximillian. Catherine Cole, and Seal Hettrick. Seal and Catherine are the two who are at the market, as a matter of fact. Unless you saw our commune car out there, a red pickup truck, which would mean they're back. It's usually parked next to our fishing boat."

"Where were these three members and yourself the evening of the murder?"

"What day was this?"

"Eight days ago."

Aronson stopped again to think for a moment. "We had a viral bug going around. Well, it was mostly me, but I didn't want to contaminate anyone. So I gave everyone the day off to rest in their cabins and relax, catching up on reading or writing or reflection."

"You didn't see or hear anyone leaving?"

"No, not a thing. Generally, although we have an open community, your cabin is your little sanctuary. People leave you alone unless you hang a green flag on the doorknob to let others know you can receive visitors."

"You wouldn't have seen the commune car disappear for several hours?"

"Each member has to check it out. Honestly, I don't recall anyone checking it out that day since I felt so poorly." Aronson took up his brisk pace again, this time taking a side trail alongside a building with a sign that read "Mail Room."

Drayco stopped long enough to peer inside, but the room was dark, and it was hard to see anything. "How does the postal mail arrive? Did Lara get many letters?"

Aronson smiled briefly. "That sign's left over from the older days when people still wrote letters. Oh, we still get some from time to time. But now it's all computers."

"So you allow modern technology here?"

"In limited amounts. There's one computer in the mail building there we all share."

"What about phones? Landlines? Cell?"

"Everyone is allowed to use the phone in the office, with all the privacy they need. But we have a strict no-cellphone policy."

"Any chance someone might sneak in one without your knowing? Or use a phone in town?"

"I doubt it, Mr. Drayco. I trust my people to do the right thing."

As they started walking again, Drayco asked, "Would it be possible to see Lara's residence? Not to disturb anything, just to look."

Aronson changed course, and they headed for a small grouping of cabins. They were separated by several feet and various plantings of marsh elder, swamp rose, and some fragrant bee balm with its faint aroma of oregano and mint. When they ducked inside their target cabin, it was mostly empty, save for some basic furniture.

Aronson flipped on a light switch, revealing that the unit was somewhat primitive with oak hardwood floors and log beam walls, but it also had all the creature comforts of a fireplace, bathroom, and kitchenette. A one-bedroom even as rustic as this would rent in the District for three grand per month.

Aronson explained, "These were once dormitory-style with outdoor toilets. We've since upgraded to single units with indoor plumbing. The smaller units used to be bunk beds with two per unit. Now, it's one person to a cabin."

"I see she didn't leave any belongings behind. Hard to tell someone lived here."

"We clean each unit thoroughly when a resident leaves." Aronson's face took on a softer tone. "Although I'd hoped she would change her mind."

He stood there with an unfocused gaze for a moment, but then he tapped his forehead and said, "Oh, I almost forgot a member, how could I?"

"Current?"

"I'm not sure."

Drayco paused his examination of the door lock. "I don't understand."

"One of our commune members disappeared. Ivon Leddon."

"When?"

"Eight months ago. We contacted his family, but they haven't heard from him."

"What were the circumstances of his disappearance?"

"He was dropped off in town to do some shopping in the bookstore. But at some point he left, according to the staff. And when the appointed time came to pick him up, no sign of him."

"Did you file a police report?"

"Yes, of course. We left his apartment intact in case he returns. I felt it was the least we could do."

"Would it be possible to see his cabin, too?"

Aronson nodded and headed out the door. Drayco started to follow, but light from the window glinting off something shiny caught his attention. He stopped long enough to rescue a small crystal wedged into a crack in a floorboard. He stuffed it into his pocket and gave one last look at the cheerless, deserted space.

Aronson guided Drayco to a nearby cabin that was a clone of Lara's, except this one still had hints of the missing resident. A cedar chest sat topped with a red-and-purple throw blanket, and some books lay on a table waiting to be read, bookended by an unopened bottle of Scotch. Drayco picked up the books to scan the titles. Nothing on criminal codes this time. Mostly westerns and a biography of Teddy Roosevelt.

Drayco's companion started picking at his cuticles again. "We truly miss Ivon. He only arrived five years ago. Was quite the handyman. Could build anything, fix anything. Some of the others we've lost, well, I may not have missed as much."

"Did you select members based on their specific talents or experience?"

"Not really. But I make it clear everyone has to share the load. That's part of our mission here."

"And the other parts of that mission? What do they entail?"

"We commit to a vow of poverty. The lure of money is what leads to most sins. I also have everyone sign a vow of integrity, a statement to adhere to an honor code. It's based on those at the military academies. Along the lines of 'a cadet will not lie, cheat, steal, or tolerate those who do.'"

When they exited Leddon's cabin, the director said, "I have to get back, but the brewery is over in that direction there. It's a bit of a hike. Feel free to chat with Max, our brewmeister. Makes the best brown ale I've ever tasted, bar none."

Before they parted ways, Drayco asked, "Mind if I take your photo? For my files."

Aronson gave him a funny look but agreed, and Drayco used his cellphone camera to take the snap. After Aronson left, Drayco looked around, making a mental map of the grounds' layout. He spotted a cellphone tower in the distance—that must be new since the Eastern Shore was notorious for its bad cell reception. The tower stood as a bit of a contrast between that modern technology and this place with its stated simple, low-tech goals. It was a techno-monster rising up out of the pristine swamp.

Well, that cellphone tower might come in handy. Drayco used his phone to do some quick research on Maximillian "Max" McCaffin. Most of what he found was old, dating back twenty-five years, but it was interesting all the same.

Aronson was right about the brewery being a "bit of a hike." Probably a quarter of a mile from the main office. Drayco headed into the brewery where the purpose of the place was immediately evident. Different-sized silver kettles and a large copper vat dominated the space, and rows of bottles filled with amber liquid perched on a table. A malty-sweet aroma permeated the air.

He headed toward a man bent over a gauge on one of the kettles and asked, "Max McCaffin?"

The man straightened up, and Drayco got a good look at him. Around the same age as the director, but much more like a Wall Street type, or what Hollywood would cast in the role, except maybe for the neatly trimmed beard. He ran a hand through his more-black-than-gray hair with a look of exasperation. "Who are you?"

"Scott Drayco. Director Aronson sent me your way. I work for an attorney defending the man accused in the death of a former member here. Lara Davidenko."

McCaffin frowned. "Death? What death?"

"Lara was murdered eight days ago. In McLean, Virginia."

McCaffin squeezed his eyes shut for a moment. "Very sorry to hear that. Gordon was afraid something might happen to her if she departed Gaufrid. She was naïve in some ways."

"Did she seem fearful or worried before she left?"

"Not to me. But maybe I mistook fear for boredom and restlessness. I'm sorry I was so dense." McCaffin lowered himself slowly on top of a small barrel near him. His face was ruddy, and even from that distance, he smelled like beer. Hazards of the trade, perhaps.

"Did she help out here, in the brewery?"

"No, she mostly helped Gordon with the bookkeeping and sales of our beer, food, and crafts. Since I lost Daven to the monastery, I've hoped we'd get new members soon so I could add another assistant. A bit rough doing it all by myself, you know."

Drayco ran his hand along the copper vat, which was surprisingly cool to the touch. "Do you make a lot of beer here?"

"It's a main source of revenue for the farm."

"Looks complicated."

"Pretty simple. Mill the grain into a grist. Auger the grist into the mash-lauter tun and combine it with hot water to make mash. Then the sparging where the sugary wort is drawn off the bottom of the tank and into the brew kettle. Then it's chilled and put into a unitank fermenter."

Drayco smiled briefly. "Sure, simple." He pointed to some large, round wooden barrels at the end of the room. "Those are impressive."

"You could live in one of these babies. They have to be big to brew enough beer at one time and let it age." Max waved at a forklift. "Only way I can move those around. All of us here are a bit long in the tooth, you know. I guess Seal is the youngest, and he's fifty-five."

"I understand a little over a week ago, a virus hit the farm."

"Yeah, that was fun. Gordon had it the worst."

"What did you do with all that downtime?"

"I never have downtime. Worked in the brewery, did some reading. I'm rather fond of science fiction."

"Did you take the car out?"

"I didn't go anywhere, no."

"Know of anyone who did?"

"The parking area is farthest from my cabin. So I wouldn't necessarily have heard it."

Drayco picked up a pH meter to inspect. "I'm curious why ex-Wall Street types would give up all the adrenaline rush to come brew beer. What led you here?"

"One gets tired of the rat race, eventually. Some sooner than others, you know."

Drayco had a fleeting thought he should get a ten-spot for every time the man said, "You know." Drayco asked, "You worked for Theunissen Trading, isn't that right?"

"Yes, although it feels like another lifetime ago."

"I read that Theunissen was fined for environmental infractions and fraud. That is, one of the company's subsidiaries, Davos Electroplating Services."

Max nodded vigorously. "Yes, it's true. I didn't know about it at the time. But I still felt responsible. The ol' buck-stops-here thing. To be honest, maybe that's the reason I came to this place. Part guilt, part shame, part penance. Hell, we all have bad things in our past we want to atone for."

"You're referring to the members here?"

"Catherine's former bosses were convicted for defrauding investors. Seal's company was busted by the SEC for insider trading.

Gordon's was accused of securities violations. Daven's for money laundering. We all wanted to leave those excesses behind."

"Does this 'penance' help your conscience any?"

"Some. And I help out with environmental causes that come my way. When an ex-member, Boyce Hershorn, ran for congress, I made him promise to make the environment a priority of his platform."

Drayco shifted one foot which threw him off balance. When he put his hand behind him, he almost launched a small unopened box into the air. He slid it back into the center of the table and studied it.

"This is from the Chesapeake Fish & Shell Company." He read the ingredients. "Pickled club crab meat. Is that a new type of beer flavoring I'm unaware of?"

Max grimaced. "Gordon is obsessed with those things. Gives them as gifts. I don't want to hurt his feelings, but I can't stomach the stuff."

"I hear you. Can't tolerate pickled anything, myself. Except for maybe pickles. I'm curious, though, how did you end up as the Gaufrid's brewmeister?"

"Came by it honest, I guess." Max chuckled. "My great-granddad brewed illegal hooch during Prohibition."

Drayco thanked Max for his time and headed toward Aronson's office, but not before taking one quick cellphone photo of Max against the beer barrels. Once back at the office, he poked his head in and said, "I was wondering about family visits? You mentioned weekend visitors. Are they allowed at other times?"

"Allowed, but discouraged. It's best if the members go to the family instead of vice versa. Max makes monthly visits to his ailing mother the last weekend of each month. Over in Manassas, I think. Gets dropped off at the bus station in Temperanceville. We have a commune account there. Seal has a daughter, though I don't believe he sees her much. But he takes the occasional weekend off, too. Catherine never married. Her remaining family is up in Maine now. She doesn't visit them or mention them much. And me, well, these people *are* my family."

"Do you conduct background research into members before they join?"

"Some. Due diligence and all."

"How much information and pre-qualification do you require? Signed vows? NDAs? Receipts? Résumés?"

"I do ask for references, but that's it. Mostly I take them at their word."

"What about Lara Davidenko? Who were her references?"

Aronson walked over to a filing cabinet. "That was years ago, of course. I only keep the application and any legal details they want. Like wills or next of kin." He held up a folder. "They're slim files, Mr. Drayco."

He flipped through the folder and said, "No next of kin. But she listed an attorney, Stuart Wissler. I have here in my notes that I called him and checked her out. Since she was a victim and not a Wall Street staffer, I didn't think it odd she had few references."

"Thanks, I appreciate your help. Looks like Seal and Catherine are still at the market. Perhaps I could come back another time and chat with them, too?"

"Of course. If it will help find justice for Lara."

It wasn't that Drayco minded hanging out longer on the shore. Especially since he'd made plans to stay in at the Lazy Crab B&B, where Maida Jepson was making her famous coconut and crabmeat soufflé for supper. But he'd hoped to catch all the members at the same time to avoid contaminating their stories.

As he climbed into his car, he filed away his encounters with two of those members, Max and Gordon. Gordon came across as abrupt at times but with flashes of shrewdness. He was a little defensive, but that was understandable if you saw your life's work fading slowly away and wondered if it had "all been worthwhile."

The beer supply truck was gone, so it was only Drayco's car and the little boat in the lot now. He didn't crank up the engine right away, using his phone to take a few more photos of the place. Despite what Lara had told her Maryland neighbor, Jilliana, Gaufrid Farm didn't exactly give off the air of a Synanon-like compound.

Why did Lara Davidenko really retreat here? She'd told Gordon Aronson some story about being trafficked by a Wall Street type. But she'd told Harry she needed a green card to avoid being sent back to her home country, where she faced sexual abuse. She'd told Jilliana that she wanted to move to another country for a clean start, but she wasn't sure she'd be safe anywhere. Which was true? Or was it all lies?

And how did Harry Dickerman manage to wind up in the middle of all of this? Was Darcie also involved, after all?

He cranked up the engine but didn't want to head straight back to the B&B. Since it was the weekend, Nelia Tyler might be in town for her part-time deputy duties. Drayco briefly toyed with the idea of giving her a call but decided against it. The last thing he wanted was to put more pressure on her than she already had. Still, he couldn't help wondering if she was okay. Maybe he should drive by her apartment to check?

Feeling a little stalkerish, he took the long way back to the B&B and slowed down when he passed her address. The light in the apartment was on, so that was a good sign, wasn't it? He pulled over long enough to tap out a text, *I'm in town for a few days. Let me know if you need anything.* But right before he sent it, he pressed delete. She was busy, he was busy, and it wasn't a good time. Would it ever be the right time? He shook his head and continued on his way.

10

Sunday, September 20

If Drayco had hoped to be able to question the remaining two Gaufrid Farm members as soon as possible, he was sorely disappointed. Not only did Director Gordon Aronson tell him Sunday wasn't a good day for a revisit, Monday was also off the list due to some inventory work the staff had to do.

Fair enough, then. He'd have to suffer being pampered by Maida and Major Jepson, the proprietors of the Lazy Crab B&B where he was staying, for another day or so. They'd insisted he take his usual room, not that it had taken much encouragement on their part. Everything was remarkably, wonderfully the same as always. Though he was amused to spy the addition of a small bust of J.S. Bach on the dresser. Somehow, he doubted Maida kept it there for her other guests.

To return the favor, he played some of her favorite Bach pieces, "Jesu, Joy of Man's Desiring" and the Italian Concerto, on their Chickering piano. He was relieved to note that most of the darkness and depression he'd felt when he played the instrument three months ago had faded. This piano gem deserved better.

Fortified with Bach, some brunch brie-bites and mimosas, and lively conversation with the Jepsons, he went out for a drive. He might not be able to question the other commune members yet, but that didn't mean he had to twiddle his thumbs the whole time.

At first, he wasn't sure where he was headed, planning on taking in some seaside views at Kiptopeke State Park or Assateague. But, as he sat staring at the marshy channels and inlets from the small parking lot of Powhatan Park, or as the locals called it, "Rainbow's End Park," he couldn't get Gaufrid Farm off his mind.

He cranked up the engine and pointed the car north toward the Virginia and Maryland border, staying on U.S. 13 until he veered off onto a gravel side road ironically named Broadway Boulevard. His GPS told him it ran around the farm's perimeter, at least the parts that were developed enough for human habitation.

When he spied a man in denim coveralls standing next to a tractor on the side of the road, Drayco pulled over and rolled down his window. "Nice day today."

The man rubbed his beard, assessing Drayco. Seemingly satisfied his visitor wasn't a threat, he replied, "Could use some rain."

"I hadn't heard about a drought."

"Not a drought. Just could use some rain."

As the man stood there with his head tilted to one side, still wary, Drayco said, "I'm staying at the Lazy Crab B&B down in Cape Unity. Thought I'd take a drive around the shore. Find one of those you-pick farms, although I guess this time of year it's all pumpkins and corn mazes."

The other man grinned, pegging Drayco as a city-slicker. "If you head back down south, you'll find some blueberries and blackberries. Maybe pumpkins and sweet potatoes, too. There are signs. Can't miss 'em."

"I saw one sign, Gaufrid Farm. Do they have you-picks?"

The man paused before replying. "They sell produce in the area here and there. Don't think they allow visitors much. Keep to themselves."

"Big family farm, then?"

"Family?" The man laughed. "Not family, no, sir. Well, not related, I don't think."

"Ah. A big corporate farm, then."

The man shook his head. "Not that neither. Don't really know what goes on there. Don't really want to know. The police have been out a couple times." He hurried to add, "Not that it's some big crime ring or nothing. One of their people went missing, I gather."

"I see. That's a shame. Hope this person turns up soon."

"Yep. My wife thinks the fellow got tired of the place and up and left on his own. A runaway farmer." He laughed again.

Drayco smiled. "It's certainly hard work."

"'Tis. Can't say I haven't considered chucking it all, myself. But this farm's been in my family for four generations. You don't just throw that away like trash."

"Who owns this Gaufrid Farm, then? Are they good neighbors?"

"I think they call it a 'commune,' whatever that means. Maybe members are part owners, I reckon. And they're good neighbors, meaning quiet and don't cause no trouble."

"It's ringing a bell now. A B&B staffer mentioned running into one of those members once. A woman. Lara David something." Drayco crafted a silent apology to Maida for the little white lie and referring to her as a "staffer." He added, "I think that was the name. An attractive blonde woman. European-looking."

"Attractive blonde, you say? My wife would kill me for admitting this. But I think I saw her at Graves Grocery. Outside Onley a ways. She sold 'em a batch of those heirloom tomatoes popular as of late. They're harder to grow."

"I'll check if Graves has some of those. And thanks for the tip."

The farmer leaned on his tractor. "That blonde woman. She smiled and all and seemed pleasant enough. But I don't know. There was something off. Like a skittish calf sensing it's on the veal truck."

The man kept eyeing the tractor and started to fidget, so Drayco said he was going to drive around a little more and continued down the "boulevard." According to the survey maps Drayco got copies of, coupled with the GPS, he was pretty sure when he arrived at one of the back borders of the Gaufrid property.

The boundary was evident from wire fencing, and as Drayco hopped out to take a closer look, he saw it was electric-wire fencing. Why would that be necessary? Few neighbors lived nearby, save for fellow farmers and the usual errant cow or a wolf or coyote. Not that the latter would be stopped by fencing since they'd just dig a trench underneath.

As he continued his drive, the road curved to the west toward the Pokomoke River. It wasn't hard to notice how close he was getting, thanks to the tributaries snaking through the landscape. Astronaut Buzz Aldrin had called the moon's surface "magnificent desolation," something Drayco was reminded of whenever he surveyed the flat, zig-zagging marshlands on the Eastern Shore. Most were still relatively pristine, unsuitable for infill or development.

The entire peninsula was mostly flat farmland in various shades of green cut through by small tributaries from the river's murky waters, which eventually dumped into the Chesapeake Bay. If you wanted a place without distractions, say, for a commune retreat, this filled the bill.

A look in his rearview mirror told him he had company. It was a black pickup typical in the area, but even for a local, the truck seemed to be going a little too fast. He pulled over on the narrow road to let the vehicle pass while he studied his survey maps. But the sound of the truck peeling out made him look up just as it whipped around and the bed made contact with Drayco's car. With a crunching sound, his car slipped off the road and onto the marshy bank where it teetered toward the water before righting itself again.

As the pickup truck rapidly disappeared the way it had come, kicking up a cloud of dust and gravel in its wake, Drayco opened the door to slide out of his car and check on any damage. That's when he realized the right side of the car was lower than the left. That couldn't be good.

After slowly exiting the car, he checked the driver's side, which showed no damage. Then he went around and saw the back right wheel was on a soft patch of muddy soil and in danger of sliding farther down into the water.

There was no way he was going to let his beloved Starfire fall into the marsh. He could try to simply drive the car to drier ground and safety, but since it was a rear-wheel drive, getting enough traction to ease out of the goo would be tricky. And it could make matters far, far worse.

The water didn't look too deep close to shore, and he eyed a thin, flat rock on a mini-island a meter away. But to get there meant doing a little wading. He slid down the bank to test out the silty bottom and took a tentative step. Unfortunately for him, the silty bottom moved, and he quickly sank into the murky muck up to his knees.

Cursing the renegade truck under his breath, he waded out to the mini-island to grab the flat rock. Then he waded back to the Starfire to scoop out a tiny indentation under the wheel where he placed his prize. After examining his work for any flaws, he used some water to wash off his pants legs.

He didn't want to get inside the car dripping wet, so he thanked the weather gods for the sunny day and sat on the grass waiting for his pants to dry. That gave him time to think about the truck—a simple careless driver? Probably not a drunk driver this early in the day. He'd only had a microsecond look at the face of the man and one letter from the license plate, "H." Was it mere carelessness or an intentional hit?

He also had time to consider his new farmer-friend's description of Lara Davidenko and how something was "off" about her. Skittish, maybe a little fearful, he'd said. But again, why? And what had scared her so much that she turned to the "only man who'd ever showed her any real kindness," according to Lara's Maryland neighbor?

After spending a half-hour in the sun, his pants felt dry enough to keep from drenching the car, so he gingerly climbed into the driver's seat and cranked up the engine. When he turned the wheel, he felt a little slip, but his rock seemed to be holding. With an extra-gentle nudge, all four wheels of the Starfire were back on the road again.

Thinking he'd had enough of Gaufrid Farm for the time being, he called up the address for the grocery store the farmer mentioned and got there in fifteen minutes. It had the look of a stereotypical country store outside, with a wooden porch and hand-painted lettering on the windows. When he walked inside, there were even baskets of penny candy stacked in rows across the front.

Drayco greeted the freckled, crimson-haired clerk with a smile. "A friend of mine tells me that Gaufrid Farm—well, one of the folks who works there—sells heirloom tomatoes in this store. I don't suppose you have any?"

The man flipped open a ledger book and studied it. "Haven't got any of those in recently. Think they brought in some squash, though."

Drayco didn't have to feign disappointment because he really wanted to taste those tomatoes. "The worker who sold the tomatoes, I think her name was Lara. An attractive blonde."

"I remember her. How could you not? You'd think she was a model, not a farmhand. Sure didn't mind her coming in, I tell you. Kind of miss those visits."

"She doesn't come anymore?"

"Hasn't for a while now. Hope I didn't make her feel uncomfortable by staring."

Drayco asked for some of the Gaufrid Farm squash, which the clerk dutifully snagged for him, adding, "Maybe she finally ditched the farm to go be a model. Sure didn't seem happy doing what she was doing."

"No?"

"I have a sister who makes a career out of being miserable, so I know what it looks like. And this Lara woman had her beat by a mile."

"She never came with any friends?"

"Just herself. And I didn't see a ring." The clerk uttered an embarrassed laugh.

Drayco paid for the squash and hoped Maida wouldn't mind him bringing it to her unasked. With her culinary skills, she'd whip up some dish he'd never heard of that made him actually like squash.

As he drove back to the Lazy Crab, he kept a sharp eye out for the truck or any other suspect drivers, but traffic was sparse today. What were the chances he'd merely encountered a road-raging truck down a lonely stretch of road in the middle of nowhere? He wasn't a Vegas gambler, but he'd put those odds around five to one against. What did that mean, then? He already had a stalker? He hadn't even been on the shore for very long.

As he passed a stand of butterflyweed with traces of their orangeish-yellow flowers, he recalled Maida saying they attracted migrating Monarchs. Lara had done some migrating, herself, from Russia to New York to Gaufrid to Maryland. Butterflies traveled by survival instincts coded millennia ago into their genetic makeup. And Lara? Was it really survival instinct that drove her, too? And why, if she'd found refuge at Gaufrid, did she suddenly up and leave that refuge after all this time?

As if in reply, a monarch flitted in front of Drayco's car, just missing the windshield. A narrow escape, but it would live another day. Sadly for Lara, whatever her motivations, her course had lured her into the path of an unstoppable tragedy from which there would be no escape.

Instinct, inculcation, inevitability—all roads and pathways of life led to the same end, no matter where you started or what choices you made. But despite that, life still had meaning and still mattered. The butterfly's life, Harry's, Lara's. Maybe the trick of living was to soar as high as you could and never look down.

11

Monday, September 21

After a decent night's sleep—except for the nightmare he'd had of being trapped in the Starfire as it sank into the water—Drayco sought out Maida in the kitchen, where she promptly handed over a large glass of something reddish-purple. He eyed it skeptically as she said, "You're not driving for a while, right?"

"I can wait an hour or so. I take it this isn't grape juice."

"A little gin, a little honey, some pomegranate nectar, and a pinch of jalapeno. I call it a Fiery Sunrise."

He grinned and took a sip. Her concoctions were tasty, if a bit. . .unorthodox. He slid into his usual chair that looked like it was carved from a ship's mast, and asked, "Have you heard much about Gaufrid Farm from the locals? It straddles the border with Maryland up in Accomack."

She sat across from him and joined him in imbibing some of the Fiery Sunrise. "When they first settled in, there were rumors it was one of those cults. The pseudo-religious kind. As the years passed, the notoriety of the place faded into this group of hippy types who kept to themselves."

"No hints of crime? Shady business practices?"

"None that I've heard. You should check with Sheriff Sailor."

She took a sip of her drink, wrinkled her nose, and got up long enough to add another jalapeno. "Major bumped into a member at the hardware store and struck up a conversation. He said the fellow seemed like a regular middle-aged farmer."

"Did this farmer give his name?"

"I don't think he did, no."

Drayco looked around for Maida's husband. "And where is the Major right now?"

"Don't laugh, but he's at the hardware store. Needed some wood to patch the fence in back."

Drayco couldn't help but laugh. "Timing is everything, right?"

After finishing the drink, he decided to take an hour for his blood alcohol content to settle down so one of the sheriff's deputies wouldn't arrest him for drunk driving. What better way than by noodling around on the Jepsons' Chickering piano? Nothing too taxing. A little soothing Debussy was just what he needed, the Arabesque No. 1.

When he felt sufficiently "sober," he headed for the former fish processing plant bought by the county a decade ago and renovated for the sheriff and occupants of the jail cells. The minute he entered Sheriff Sailor's office, the man greeted him with, "You look like you could use a drink."

"Well, I just—"

"Come with me."

Feeling a little bemused, Drayco followed along behind the lawman, who led him into a room at the back of the facility where Drayco hadn't been before. One corner sported a beverage station with juice and sports drinks, while an array of weight machines and benches filled the rest of the space.

"You put in a gym?"

"Who wants fat cops? This used to be an old storage closet. Funny how much room you can get when you throw out a lot of worthless junk."

"How can you afford all of this on your shrinking budget? I know the county board would not approve."

"Got it all second hand from an estate sale. The daughter of the man who died practically gave it all away, she wanted it out of the house so badly."

"If her *late* father used that gear, it didn't exactly help him, did it?"

"Car accident got him, actually. Drunk driver."

"Ah." Drayco was extra glad he'd waited to drive after Maida's welcome-back potion.

Sailor went to a small refrigerator and pulled out two bottles, handing one over. When Drayco saw what it was, he crowed, "I've made you a convert."

The sheriff opened his bottle of Manhattan Special and took a swig. "Just helps keep Limping Mike's Bait Shop in business."

"Uh huh. You'd never admit I got you addicted to these." Drayco maneuvered around a rack of weights to a machine he didn't recognize. "What does this do?"

"Bicep curls. Like this." Sailor sat down and worked the contraption to demonstrate. When he hopped up, he said, "Care to try it?"

Drayco put his drink down and slid into the vacated seat and pumped the handles. Sailor studied him. "That seems too easy for you." He changed the pin to add in twenty more pounds of weights and said, "Try that."

Drayco did, and Sailor's jaw dropped. "It thought you had a gimpy arm. That's more than I can lift."

"My lower right arm was damaged, but the biceps are fine, thanks."

"How do you keep in shape, then?"

"Playing the piano requires more muscle than you'd think. Even as little as I'm able to do."

"Oh?"

Drayco rubbed the back of his head. "I have a set of weights at home, too. Part of my physical therapy."

"Aha! I knew it."

Sailor leaned against an elliptical machine. "So, you're interested in Gaufrid Farm up the road, eh? Least, that's what you said on the phone."

"It's a murder case I took on for Benny Baskin."

"Who's the client?"

"Harry Dickerman."

Sailor narrowed his eyes. "Isn't that the guy who's marrying Darcie Squier?"

Drayco pursed his lips. *This again.* "Yeah, that would be the one."

The sheriff stared at him. "Hoo boy. That must be awkward. But what does Gaufrid Farm have to do with it?"

"The murder victim was a resident there until recently."

"You don't say? Never had any dealings with the place. I knew of the commune, sure. Checked them out, out of curiosity. But since there's never been trouble, and it's not in my county, it wasn't on my radar."

"What about police in other nearby counties?"

"After your call, I asked my counterparts in Accomack and Maryland to see if they had anything of interest. Got a lot of blanks. No crime reports. No complaints. The commune paid their taxes, and no one ever saw much of the members."

"Nothing of interest whatsoever? After twenty-five years?"

"They must be tissue farmers, they keep their noses so clean."

"Maida said there were rumors at first it was a cult."

"You know how people are. If you don't belong to their particular religion and worship in their exact way, you're suspect, right? But I have no info these folk are religious of any stripe. Whatsoever."

More disappointments and dead ends. Frustrated, Drayco pressed Sailor, "Does anyone have a grudge against the commune locally? Property disputes? Anything to want to force them to leave?"

"People with a beef would just trash the place. Or burn it down. Or try some legal maneuver. Murder's a tad extreme, don't you think?"

"And the missing man I told you about, Ivon Leddon? Any reports?"

Sailor gulped down his coffee-soda and went to get another one. "I guess that's one odd thing. Plus, a while ago, some equipment went missing. Mostly brewery stuff, but you'd expect that. Beer lovers, maybe. Copper thieves, too."

"The farm's director, Gordon Aronson, mentioned he'd filed a missing person report with the police about Leddon."

"Sheriff Quarles in Accomack said as much, too. Since he saw no signs of foul play, it went straight to the back burner."

"Seems odd he'd disappear, and eight months later, another member leaves and is murdered." Drayco stood up to inspect a rowing machine. Who needed that when they could have the real thing less than a mile away?

"Yep, kinda glad it's your case and not mine."

"Did you know another ex-member is now a U.S. Representative?"

"Oh, swell. I'm doubly glad this isn't my bailiwick, then. 'Cause I don't want to get anywhere near a Congress Critter in any investigation. It's all yours, Drayco. Be my guest."

"Wuss."

"When it comes to getting political dirt on my shoes, you bet. You staying with the Jepsons?"

"Naturally."

Sailor folded his arms across his chest. "Trying to get Darcie's fiancé off the hook, my my. How's that working for you?"

"It's a job. The victim hires Benny Baskin, Benny pays me, I put the money in my bank account. That's all it is."

Sailor didn't push any further. But the man had his fair share of investigating difficult personal cases. . .like Darcie's ex-husband, Randolph Squier, now serving time for embezzlement.

Sailor chugged more of his second soda. "Ivon Leddon. Sounds Russian, doesn't it?"

"Funny you should say that. The victim's name was Lara Davidenko."

"Also Russian?"

"Allegedly."

"Isn't that a coinkydink?"

"Maybe."

Drayco wasn't matching the sheriff gulp for gulp with his drink, something he tried to rectify before saying, "Nelia Tyler should stick around this place. Look at these perks. Biceps and Manhattan Specials."

Sailor glared at him. "You don't wear sarcasm well. And for your information, Nelia gets a chance to use this gym when she's here on weekends."

An image of Nelia working out on the machines made Drayco's thoughts stray a little too far into unprofessional territory, so he hurriedly asked, "How about Mrs. Sheriff? You let family use it, too, right?"

"My wife says she gets plenty of arm lifts doing the laundry. And she's not terribly fond of sweat."

"Women glisten, not sweat, right?"

Sailor chuckled. "Better not tell her that. She'll show you just how strong those laundry-arms are when she boxes your sexist ears."

Deputy Wesley Giles poked his head into the room and waved at Drayco before addressing Sailor. "Got Judge Williams on the phone. It's the Exeter case again."

"Tell the good judge not to get his knickers in a twist. I'll call him back."

Giles bit his lip. "He says it's urgent."

"It's always urgent with him. Sure, fine, I'll take the call." Sailor nodded at Drayco. "If you're going to be in town a day or two, we can grab some grub at the Island View Restaurant."

"Seafood Hut not rebuilt yet?"

"Might take another few months, they said."

"Island View, it is." Drayco watched him go and tried the bicep machine one more time. Too bad they didn't have machines that reversed nerve damage in a mangled arm. If they did, he'd be more confident taking on that recital at UMD.

Or maybe he'd cancel the whole thing. No one wanted to hear a washed-up pianist, certainly not one who made his living culling through the flotsam and jetsam of shipwrecked souls.

He shook his head. What was it with all the marine metaphors? That's what happened whenever he was here—the bay, the sea, the marshes—everything ended up back at the shore. Even his thoughts, his words. . .and his nightmares.

Sailor had wondered about a possible Russian connection, and he wasn't the only one. Though the sheriff hadn't helped with intel regarding Ivon Leddon, the missing-man link was intriguing. Another Russian tie to the case, like the victim and Alisa Saber's missing Russian mother? Maybe the man had disappeared intentionally to become the "invisible man," making it easier to kill Lara Davidenko. But why wait until she went to see Harry? There was plenty of time to do the deed after Lara moved to Maryland.

At any rate, the commune seemed a long shot. The members were on downtime due to a virus the day Lara was murdered, and it would be hard to travel to McLean, commit murder, and get back, given they all used one communal car. But the Leddon-as-killer scenario? That was far more intriguing.

12

Something Gordon Aronson had said at the commune prompted Drayco to take a trip to the Oyster Creek Marina. Being around boats again made him think of Darcie taking jaunts on Harry Dickerman's yacht moored at the Occoquan Marina, farther north and closer to D.C. He doubted any of these Oyster Creek craft were quite as big as that.

Thanks to a tip from Sailor, Drayco parked in the small gravel lot and strolled along the wooden boardwalk toward one particular mooring slip. The white houseboat tied up there had the name *Phobos* painted on the stern.

When he got closer, he saw the boat was about forty feet long with decks in the front and back and a platform on top with rails. A man wearing a navy beanie, teashade glasses, and a golden fishhook earring that matched his golden beard ducked outside and planted himself on the deck.

The man stared at his visitor for a moment, then said. "Long time, no see, private cop."

"Still doing marine salvage?"

Dennis Frischman grunted. "Pays the bills."

Drayco waved his hand at the name on the boat's stern. "Phobos, like the Greek god or Phobos, the moon of Mars?"

"Guess it could be both. Phobos was the god of fear, right? Probably a lot of fear to go around when my salvaged-ships sank. I also read something about that moon not too long ago. More of a rubble pile that's slowly falling apart. Either way, pretty apt."

"There's talk about mining moons and asteroids. That could be your next venture."

The other man laughed briefly. "I'll stick with undersea treasures, thank you very much. But, I doubt you're here to buy any recovered antique Fresnel lights or compasses and binnacles."

"It has more to do with your past life."

Frischman squinted at him in the afternoon sun. "Another suicide-not-suicide?"

"Murder this time."

The man waved Drayco into the houseboat, which was surprisingly homey. Full galley with stove, sink, and mini-fridge, a sofa bed, even a window AC unit. He pointed at the sofa bed for Drayco while he sat on a blue fold-down chair. "What made you think of me and my past life?"

"I was up at Gaufrid Farm yesterday, the commune for Wall Street escapees. The director there hoped he'd be able to add more members soon, noting there's always an economic downturn around the corner."

Frischman nodded. "He's right, there."

"He also remarked that joining the commune is better than throwing yourself off a building. Seems like that's the same thing you told me back in the summer. At least, the throwing-yourself-off-a-building part."

Frischman stroked his beard. "Might recall having said something like that. This murder. . .one of the Gaufrid guys involved?"

"One of their ex-members, Lara Davidenko."

"Ah. Former, though, you say?"

"Left three weeks ago, moved into an apartment in Maryland, looked up her fake ex-husband, and was killed in his house."

Frischman whistled. "That's a nasty piece of work. What's all this about a fake ex-husband?"

"She married him under the pretense of getting a green card to avoid trafficking. Then stole his money and disappeared, apparently to the commune."

"What country? Her name sounds Eastern European."

"Russia."

"Not surprised." Frischman grabbed a cigarette from a pack in his pocket. After he lit it and made one of his perfect smoke rings, he added, "It's most often Asian or Eastern European girls."

"Are you telling me there are gangs of Wall Street traders who run trafficking rings on the side?"

"More like Wall Street investors who launder money from drugs and trafficking into their real estate holdings. Happens more often than people know."

Drayco reached into his wallet and pulled out a hundred-dollar bill. As he handed it over, Frischman said, "What's that for?"

"Last time I ran into you, I'd given my one remaining twenty to the bartender. I owed you a tip. Still going to the Boardwalk Bar every night to drink your teetotaling sodas?"

"Yep." Frischman got up long enough to grab a couple from the mini-fridge and hand one over.

Drayco took a few sips. Not Manhattan Special, but pretty good. "Since you're a former Wall Street player, do you associate with the Gaufrid gang?"

"I know who some of them are. By name only. But associate with them? I prefer the fish. Flounders are better company than felons."

"Funny you should mention that. Seems like several of the 'farmers' used to work for felonious companies. Money laundering, environmental crimes, insider trading, defrauding investors. A little bit of everything illegal."

"Which companies?"

Drayco rattled off the names of the businesses he'd learned from his research, and Frischman nodded. "Familiar with all of those. Illegal is standard practice on Wall Street these days. The only real crime is getting caught. Makes me skeptical about a commune for ex-Wall Street-turned-peacenik types."

"They do say twenty-one percent of corporate execs have psychopathic traits."

Frischman took a drag off his cigarette and blew some more smoke rings. "Don't know of too many from The Street with a conscience. You hand over your soul to work there. More addictive than cocaine."

"Funny. Director Aronson said the same thing, only he said heroin. And leaving cocaine—or heroin—for cooking, cleaning, and contemplating your navel doesn't seem normal?"

"Not quite." Frischman leaned back in the fold-down chair, careful not to bump into the sink next to him. "You spoke with the director, you said. I think his name is Gordon Aronson?"

"He and Max McCaffin. The only other two remaining members, Catherine Cole and Seal Hettrick, were out getting supplies."

"Yes, well, did Aronson spout some of that mission gobbledygook about personal integrity and vow of poverty?"

"To some extent."

"I'll bet our communers aren't all that poor. You checked to see if they're active traders?"

"I did. Haven't found any activity so far."

"Could set up trading through an intermediary. I heard a rumor Aronson is a millionaire."

"From trading?"

"Who knows? It ain't from a rich wife. His ex was the one who filed for divorce. The standard irreconcilable differences, although there was an affair, I think. Also standard in the finance business. The ex is remarried to the owner of a multi-million-dollar corporation."

"If what you say is true about Aronson's financial status, the commune might kick him out, and it would fall apart. Or maybe he just has a bunch of leftover bank accounts and mutual funds."

Frischman laughed. "You can take the man out of Wall Street, but you can't take Wall Street out of the man. Becomes part of your DNA."

"Even you?"

"I still trade, but it's fish and salvage parts, not little green bills—or more like bitcoin, since it's all digital trading now." He stood up. "Wanna take a nickel and dime tour of my sea palace?"

Frischman pointed around the cabin. "It's got most of the basic necessities." He opened a small closet to reveal a toilet and shower combo and then flipped open a cabinet on the wall.

Drayco expected to see books or linens, but instead, the open doors revealed a sophisticated-looking tech setup. He asked, "Satellite?"

"For phone, internet, TV, weather, and navigation."

"This place may look primitive on the outside like Gaufrid Farm, but it's not *primitive* primitive."

They headed toward the front of the houseboat where the central console was. The steering wheel and main controls seemed old-fashioned, but when Frischman flipped on the switch, a split screen panel lit up showing sonar, chartplotter, radar, and weather displays. Frischman said, "Some of this is old, some of it's new. I cobbled it all together into something that works pretty well."

Drayco pointed to a large locked box on the bow of the boat. "What's in there?"

"That's all the high-tech toys for salvage. ROV, magnetometer, and some SCUBA gear."

Drayco surveyed the cabin. "I'm envious of your setup. Except there's no place to put a piano. Maybe a digital piano."

Frischman tugged as his beanie. "Piano?"

Drayco smiled. "A hobby." He reached over to touch the screen that resembled the glass cockpit of a plane. "You said you bumped into some of the Gaufrid gang. Did you ever meet Lara Davidenko?"

"It's not a name I'm familiar with. What company did she work for?"

"She didn't. She told Aronson she wanted to stay there because she was fleeing from her evil Wall Street handler. It was the same song-and-dance she gave the ex-husband about running from sex trafficking."

Frischman peered at Drayco over his teashade glass frames. "I suppose it's not impossible. They say human trafficking is the third biggest criminal enterprise. Any wonder Wall Street would be involved?"

"You said the director, Aronson, may have a big money stash. Could he have killed the victim when she found out, since they were supposed to take a vow of poverty?"

"Again, not impossible. But I can tell when someone is fishing. And you're still in the shallow end, my friend."

"There are plenty of Wall Street-mob connections, maybe even with this case. Also not impossible?"

"Feels more like an isolated thing. The lone wolf or a small group of white-collar crooks. I mean, not many mobsters could stand living in such an out-of-the-way joint as Gaufrid Farm. They're high rollers. They like the fast life too much."

Drayco thanked Frischman for his candor, and the man called out as Drayco exited the boat, "Having worked alongside high rollers, I would caution you to watch your back. Some of those guys are not all sugar and spice."

"Thanks for the warning. Assuming you still have some connections to your former life, if you hear of anything. . ."

Frischman laughed. "Fishing is my specialty. If I catch anything, I'll fry it up for you and send it your way."

Drayco stepped out onto the dock, inhaling the fishy air. An incoming boat loaded with bikini-clad young women navigated into the slip a couple down from Frischman's houseboat. If this were a regular sight, it was obvious why the man had chosen his spot.

One of the women blew Drayco a kiss as he sprinted by. He smiled and waved, but kept heading toward his car. Not that spending a little time watching a boatful of attractive women wearing not much of anything wasn't something he'd ordinarily enjoy.

His chat with Frischman was interesting, if not as enlightening as he'd hoped. The whole trafficking thing, though—surely it had been too long for that to be a motive? His gut instinct was telling him something else was at play, like Frischman said.

Since Frischman's ordinary soda just whetted Drayco's appetite for something better, he stopped by Limping Mike's Bait Shop to grab some Manhattan Specials. After his last investigative case in these parts, he'd gone off his usual salted coffee. But that hardly meant he didn't need caffeine. And espresso sodas should do the trick.

Salted coffee. . .he had a sudden image of finding out his mother also salted her coffee when they'd reconnected seven months ago. Was it only seven months? Seemed another lifetime. Just like the eternity since she'd taught him how to play the piano when his legs couldn't reach the pedals. But her life had been a lie, making his whole childhood based on a lie.

As he headed back to the Lazy Crab, he started tapping out the rhythm of "Would I Lie to You?" on the steering wheel before realizing he was doing it. Lies, indeed. There seemed to be plenty of those going around, and Frischman was right—Drayco was still fishing in the shallow end for answers to those deeper questions.

Maybe he'd get more of his answers tomorrow when he returned to the commune. How did Frischman word it? You can take the man out of Wall Street, but you can't take Wall Street out of the man. Drayco was beginning to suspect that his initial view of the commune not having anything to do with Lara's death might have been hasty.

Two Gaufrid Farm member interviews down, two to go. He just hoped the extra delay hadn't tainted their stories, adding more reefs along his course. Or allowed them plenty of time for cover-ups. Couldn't be helped. . .though he briefly toyed with the idea of sneaking over that electric fence after dark.

13

Drayco reluctantly stopped by the Opera House on the way back to the Lazy Crab, surveying his contractor's latest work. The place reeked of fresh paint, and the stage flooring was two-thirds installed. It was painful to see the project in its half-finished state, knowing that if more funds didn't come in soon, it could all be for naught.

What would become of the piano he'd just had refurbished, the same piano played by Konstantina Klucze and other notables? He really didn't want to think about that. There were too many other ghosts to contend with.

Satisfied everything looked as expected, he headed toward the rear stage door but stopped short. After reading through old news accounts about the Cape Unity Opera House, he'd found a peculiar tradition among performers there. Before each concert, it became a sign of good luck to rub a white brick, exposed after someone broke a hole through the interior plaster wall. As legend had it, it happened during an accident with a piece of stage equipment, but the real reason didn't matter to Drayco. He wasn't about to cover it up.

He fingered the brick where it was rubbed smooth through the years. He didn't believe in superstitions, but he wouldn't turn down a little luck. Or maybe the universe could offer up some insight into Lara Davidenko—who she really was and why she ended up dead in Harry Dickerman's living room.

After giving one last look at the stage, casting a wistful eye over at the closet that harbored the piano, he left the work-in-progress with a sense of resignation. *Que será será.*

When Drayco returned to the Lazy Crab, the evening meal was still a few hours away, so he settled into his room with his laptop. He would tackle Alisa's case later, but first, he had something else in mind. Before he'd headed to the shore, he'd scanned in the old photo of Lara Davidenko from Harry via Darcie, and downloaded some de-aging software.

He spent over an hour working with the settings. After he tried several versions, discarded them, and tried several more, he finally had what he thought was a pretty good photo of what Lara looked like thirty years ago. If she *were* trafficked from Russia, it was most likely she arrived in the U.S. in her teens or twenties. Traffickers preferred the young, naïve ones, and unfortunately, there seemed to be an unlimited supply for their purposes.

Reasonably pleased with his latest version, he started looking through some old news databases he had access to, focusing primarily on those from the mid-Atlantic area. After all, this woman didn't stray far from the coast, moving only as far as Salisbury after she left the commune.

Looking at old mugshots wasn't his favorite thing to do, and after a while, he was afraid he'd get sloppy due to "photo fatigue." So he took a break to lie back on that incredibly comfortable bed the Jepsons had bought for this room.

Was Frischman correct that this case was a lone wolf rather than an organized mob link? On the other hand, Sarg would start salivating at the slightest hint of mob involvement. Drayco's former partner was a sucker for "classic" FBI-style cases.

At least two people were involved with Lara Davidenko's shady past, Lara and Stuart Wissler, and one was dead. The latter was Drayco's next target since he seemed central to this woman's past. Thanks to Benny Baskin's digging, they already knew Wissler was a fake attorney—and a fake who-knew-what-else.

With a sigh, Drayco sat up and started scrolling through the database again. He *must* have had photo fatigue because he almost missed an item as it scrolled on by. He stopped, went back a couple of pages, and stared. Grabbing his de-aged photo, he compared it to the one looking at him from the screen. The women could be twins.

The date on the arrest record was thirty years ago. That would mean his guesstimate got the age about right, since the woman in the mugshot was twenty-two at the time. The charge was for passing bad checks, but she was subsequently released. However, the name on the record wasn't Lara Davidenko. . .it was Minna Hallow, with a hometown listed as Brooksboro, Maryland. Drayco looked it up. Not far from Salisbury, with a population of under five thousand.

If this was the same person, what about her accent? Then again, Jilliana Vaughan *had* said she was surprised Lara's accent wasn't all that strong, "her being Russian and all." Well, if Lara and Minna were one and the same, it made it less likely "Lara" was Alisa Saber's mother—although that was always a long shot.

While he was in research mode, he dug into what he could find on Stuart Wissler, the shady lawyer-probably-not-a-lawyer who'd arranged Harry's marriage to Lara. Benny may have tackled the attorney part of the man's alleged background, but he hadn't gone deeper. Drayco spent another hour on that without much luck—until he decided to use a different newspaper database of old articles dating from twenty to thirty years ago.

One mentioned a "Stu" Wissler, who was rumored to be a pimp for vulnerable women, mostly foreigners and particularly eastern Europeans. He was arrested briefly but vanished while on bail. The article had a small photo of Wissler, but it was grainy—not much to go on.

Okay, so how would little Minna Hallow from small-town Maryland connect with someone like that? Was that why she'd pretended to be Russian? Maybe she wasn't running from a Wall Street abuser, but rather a pimp abuser.

Yet, Gordon Aronson said Stuart Wissler was the reference she gave Aronson before she entered the commune. So the fleeing-a-pimp motive didn't work because Wissler would have known where she was. Perhaps he was even the one who helped arrange for her to get into the commune. But again, why?

Drayco pulled out his cellphone and dialed a familiar number. The voice on the other end said, "If you're calling to invite me to fly with you again, I've been grounded. So to speak. Director Onweller wasn't amused over that whole tangle with the police thing. Like it's my fault."

"Sorry about that. Well, the grounded part. The ticking-off Onweller part on the other hand. . ."

Sarg's voice had a hint of amusement as he replied, "Don't think I'm too deep in dog doo. And you'd better *not* stop flying me places, or I'll disinherit you."

"I'm in your will?"

"I left you that ceramic chicken with the missing eye you love so much."

"Gee, thanks."

"Where are you, and what's up?"

"Eastern Shore, and I have a task for you."

Sarg snorted. "Am I your lackey now?"

"Yes. Yes, you are. Look, I need you to check with your contacts to see if you can ID the victim's name, Lara Ekaterina Davidenko, as a possible stolen identity. Might be Russian. Or somewhere in eastern Europe."

"That narrows it down. This victim of yours. . .not really Russian?"

"It's possible her real name is Minna Hallow, and she's as American as Maida's honeycrisp-and-rhubarb pie."

"I could make a few calls. *If* you promise to fly me back to Cape May for some crab cake burritos."

"Deal. I'll even pay for your meal if you'll also check up on a guy named Stuart or 'Stu' Wissler while you're at it. He might have had some actual trafficking in his background. But if so, he's mostly a ghost in the system."

"You don't ask much, do you? Yes sir, boss, I'll see what I can do, boss." Sarg added cheerfully, "Think you can throw in some steep turns and stalls while we're en route to the Cape? Found myself wishing we'd got arrested at the apartment. Methinks my life is getting a little dull."

"Polka music not cutting it for you?"

"I dragged out my Black Sabbath album the other day."

"Oh, Sarg. You need an intervention."

"Cape May. Crab cake burritos and key lime mousse pie. Don't forget, junior."

Drayco hung up and rubbed his eyes. After all that computer work, he almost wished he had some reading glasses, but he was going to put those off as long as possible. Maybe when he was eighty. Tired of research, he headed downstairs to have some interaction with Maida and Major.

She had more of the Fiery Sunrise ready for him, though considering the time of day, more of a Fiery Sunset. This time, he was glad he wasn't driving because the drink tasted twice as potent. "Am I your only guest right now, Maida?"

"Our trade falls off after Labor Day. But it'll pick back up when the leaves turn. Cape Unity in the fall with colorful foliage and the water is a pretty enticing mix. We also get more customers around Thanksgiving, believe it or not."

Drayco started to ask where Major Jepson was until he spied him in the garden, his usual spot, tending to some coastal sweet pepperbush plants. When the man looked up, Drayco waved.

Maida folded her arms as she leaned against the stove. "I was surprised to hear you're working to prove the innocence of Darcie Squier's fiancé. Has to feel odd to do that."

Drayco didn't answer right away and bit his tongue, irritated by her question. As he'd told Sheriff Sailor, a client was a client was a client. He was a professional consulting detective, not some sixteen-year-old lovesick schoolboy.

The ever-perceptive Maida refilled his glass and patted him on the shoulder. "Think I'll start calling you 'Saint Scott.'"

The irony wasn't lost on Drayco that he'd mockingly called Harry Dickerman the same thing. "But you're a Methodist lay pastor, not Catholic."

"We have our saints, too. Not officially, of course. And anyone who can take on a task like this one deserves sainthood. That, and some Cajun shrimp and grits followed by dark chocolate pecan pie."

"Hope you don't let Sheriff Sailor know you made dark chocolate pecan pie." Between Maida's cooking, Sarg's seafood obsession, and

Sheriff Sailor's pie addiction, Drayco would weigh three hundred pounds before long.

Maida took a sip of her drink. "I made a few discreet inquiries among my friends in these parts. About Gaufrid Farm. Wish I could say I uncovered something salacious. The worse thing anyone said was they were a little tired of all the beer truck supplies coming and going. The farm sells a lot of their beer."

"Have you tried it?"

"No, but I think I will after all of this."

He shook his head. "I don't know, Maida. Might make you disappear like Ivon Leddon."

"Beer that's like disappearing ink? Now I've definitely got to try some. Just for the look on Major's face."

Disappearing ink, disappearing men, disappearing mothers. He wasn't nearly as far along in the case as he'd hoped to be by this point. But if the Minna Hallow link panned out, things were looking more interesting, if also more perplexing.

He pulled out a deck of cards and shuffled several times. As hand and arm therapy went, it was an easy "device" to carry around, though he doubted any therapy was enough to get him through an entire concert. He could just see it now, playing a movement, stopping to shuffle some cards or soak his arm in warm water, and then playing another movement. *Come for the piano, stay for the circus.*

It was the luck of the draw that ended his piano career—and apparently the luck of another bad draw that ended Lara Davidenko's life. Now, if he could only figure out what dangerous game she'd been playing and why.

14

Tuesday, September 22

Drayco called ahead to Gaufrid Farm, and Gordon Aronson said the members who'd been out on an errand two days ago would finally be available. Drayco grabbed some Manhattan Special espresso sodas from Limping Mike's Bait Shop he guzzled down during the drive to the commune. It was a hazy, humid morning, the type where the sky threaded into the air below like a hot, wet blanket sucking the oxygen out of your lungs.

After parking his car and leaving the windows cracked to let out some heat, he checked in first with the director. When Drayco remarked on how isolated the commune appeared to be, Aronson nodded and said, "A reason I picked this place. Also reminded me of my grandfather's farm, God rest his soul. If God took him, that is. I have my doubts."

At Drayco's raised eyebrows, Aronson explained, "My grandfather was OCD about business. Driven, worked nonstop. Wasn't happy, though. He taught my father to be driven and work nonstop, too. And also to be unhappy."

Aronson grabbed some files from his desk and shoved them into a cabinet. "They both convinced themselves they were happy. But deep down, they were rich, miserable SOBs who didn't know what happiness really is. Then again, who does?"

"And you decided to step off that treadmill?"

"Got tired of the brass ring that's always out of reach. Or if you do catch it, you want another, then another, and can't seem to stop the chasing. For what?"

Drayco thought of Dennis Frischman's rumor that Aronson was secretly rich, but since he didn't have proof, he opted for a more oblique tactic. "I'm curious about Niles Peto's cocaine overdose. Aren't members supposed to take a vow of abstinence against such Wall Street vices like drugs. . .and money?"

Aronson picked at his cuticles. The man seemed oblivious to the habit. "We do offer counseling as needed. Sometimes it isn't enough. Cocaine is a harsh mistress, but we thought he'd got it licked."

Drayco looked above Aronson's head to a photo on the wall with a picture of the cabins. "How did you come up with the name Gaufrid?"

"From the Old High German name, *Gaufrid*, itself composed of two elements, *gawją*, meaning land, and *friþuz*, meaning peace and friendship. The land of friendship, the land of peace."

"Have you found that here after all this time?"

After a slight hesitation, Aronson said, "You might say that. But you want to speak with Catherine and Seal. You'll find Catherine near the barn." He pointed out the window toward a reddish building off to their right.

With that abrupt dismissal, Drayco strolled across the gravel-and-dirt path from the office to the barn, following the odor trail of animal excrement as he took note of the swamp azaleas nearby. Naturally occurring or transplanted? As Drayco visually swept the area nearby, he identified some fruit trees—apple, pear, cherry, peach. Walnut and pecan, too.

He noticed a red-haired woman in a pair of denim overalls bending over a chicken coop inside a fence that surrounded the barn. He approached her and asked, "Catherine Cole?"

She straightened up so fast, she almost hit her head on the coop frame. When she saw him, she relaxed. "You must be Scott Drayco. Gordon warned me you were coming this morning. He said I should cooperate." As Drayco walked through the gate, she added, "Mind if I keep working while we chat?"

"Sure." He glanced behind the fence where two cows munched on grass. "I think I can guess your role here."

"I take care of the animals and cleaning. It sounds sexist, but I had a choice of jobs. This is what I picked. I was raised on a farm, anyway. Animals love you unconditionally, don't they?"

"And they don't usually talk back. Unless they're huskies or parrots."

She smiled. "I have conversations with our animals. Jake, the horse, thinks he's a fire-breathing dragon."

"Is Jake a stallion or gelding?"

"The latter. But don't tell him."

Drayco chuckled, but a cow mooing made him curious. "Isn't it a bit hard to eat the animals you raise?"

"We consume a mostly Mediterranean diet here. Fish, eggs, cheese, nuts, fruit, veggies, grains, and beer. I never touch meat. I believe animals have souls, too. They're angels in disguise."

Drayco watched as Catherine sprinkled some corn feed into the coop and checked the nests for eggs. The wood on the coop looked new, as did the fencing around the barn. He pointed it out. "That's some good workmanship. Looks like things are well maintained here."

She picked up a few eggs from the nests and straightened back up. "Why, thank you. I made both of those."

"Guess I should have said it's nice work-woman-ship."

"I don't believe in all that non-sexist language thing. I just want to be judged by the work. I think of myself as a holistic thinker. Everything in the universe is interrelated."

"Sounds like Reiki or Shamanism."

She smiled at him. "I'll bet your spirit animal is a dragon. They offer the energy of ancient magic and primal strength."

He replied, "Would fit my last name."

She started checking the chicken nests again. "Aren't you going to ask me about Lara's murder and whether I did it or not?"

She'd just confirmed Drayco's suspicions Gordon would contaminate the others with details about Lara before Drayco could talk to them first. Maybe he should have forced the issue by showing up unannounced a day ago. "You and Lara were the last remaining female members of the commune. You must have been good friends."

Catherine paused her task for a brief moment before starting up again. "We got along well enough. She was one smart cookie. Her talents were wasted here."

"Talents?"

"She was so good at finding secrets, she should have been an investigative reporter. Won a Pulitzer, even."

"By finding secrets, are we talking blackmail material?"

Catherine's lips curled into a half-mocking smile. "She wasn't blackmailing me, if that's what you're asking."

"The others, then?"

"You'd have to ask them."

"I will. Except for the missing man, Ivon Leddon. Was he being blackmailed?"

"I doubt it. His chakras were too strong and balanced." Catherine pushed a stray hair off her cheek. "Poor Ivon."

She cut Drayco off before he could ask what she meant, "Maybe it would help more to say he was an Eagle Scout type. Kind of Nordic-looking."

Drayco tilted his head. "Chakras and spirit animals? Do you believe in the Akashic Records?"

She eased some more eggs into a waiting basket. "No, that's baloney. Just science fiction. You don't believe in that bunk, do you?"

He stopped himself from pointing out there was little difference between the Akashic Records and her philosophy. Instead, he reached into his pocket to pull out the small crystal he'd found in Lara's cabin on his first visit.

Holding it up in the air, he asked, "Do you know what this is?"

She stared at it for a moment. "Where did you find it?"

"In Lara's cabin, wedged into a floorboard."

Catherine shrugged. "A talisman. I probably dropped it there once. They bring good luck."

"It didn't seem to bring Lara good luck."

"She wasn't a believer."

"Is that a prerequisite for those to work?"

Catherine just shrugged again. It wasn't lost on him she hadn't denied the talisman being hers. Not only did she *not* deny it, she even volunteered it might be hers. Did she have nothing to hide, or was she trying to evade the question?

He said, "Getting back to Ivon and his secrets—"

"Can't imagine he'd have any secrets whatsoever." She waved a hand in the air. "I'm not sure why he felt he had to come here. Unless he was having a crisis of conscience about accidentally overcharging someone. What a crime *that* would be."

"He wasn't well-liked?"

"I don't want to give the wrong impression. He was handy to have around. And I honestly have no idea why he would up and disappear like that."

"Aren't you afraid something bad has happened to him?"

"Bad? You mean because of Lara's murder? No, must be a coincidence."

Catherine paused a moment, then continued, "We're hardly a den of thieves here. Just lapsed traders and dealers who still had enough of their souls intact to try to avoid eternal fire and damnation."

"Then why did you imply blackmail? Especially if, as you say, you weren't a victim?"

"I overheard Lara talking with two former members using terms that certainly sounded like blackmail. Both are dead now." She rushed to add, "Natural causes."

"Did Lara mention the name Minna Hallow?"

Catherine scrunched up her forehead. "Minna Hallow. No, doesn't ring a bell."

Drayco didn't yet have concrete evidence about Lara's true identity, and Catherine's denial seemed genuine, so he changed the subject. "Director Aronson said commune members do see family from time to time. But what about husbands? Wives?"

"Sex, you mean?" She smiled at him. "Let's just say we have an open commune. Live and let live. What goes on inside a cabin stays inside a cabin."

"Sounds like a soap-opera-in-waiting."

"We all hooked up with each other at some point."

"Even you and Lara?"

She winked at Drayco. "A few times. I swing whichever way the wind blows." She bit her lip as she added with a tinge of bitterness in her voice, "I think Gordon was secretly in love with Lara, though."

"Did she return those feelings?"

"Unfortunately, no. He was pretty angry when she left. Has this controlling vibe at times."

"How upset was he?"

Catherine stuck out her chin. "Just upset. I really didn't get it. A good-looking man like him, why. . ." She caught herself. "Whatever. It's not my business."

"Perhaps Gordon was jealous when he found out Lara was going to see her ex-husband?"

"She never mentioned this Harry fellow before. We didn't know she'd been married until you told Gordon."

"Were you also upset Lara left?"

Catherine took a deep breath. "Her aura had turned into a muddy gray color, which signifies fear or depression or confusion. It made me think she'd had enough of the simple life. Wanted to get out and party again. We've lost several members that way."

"Gordon told me about a few of those members."

She wiped her hand on her sleeve to rid herself of some clinging bits of straw. "Guess I can't blame them. That micro-celebrity lifestyle is a big lure." She peered at him. "Speaking of celebrity, your name sounds familiar."

"Does it?"

He was expecting her to mention his piano days, but she surprised him. "You're the new owner of that Opera House down in Cape Unity, right?"

So Catherine kept up with the news, too. "Fortunately or unfortunately, depending on how you look at it."

She smiled and then nodded at the barn. "Look, I've got to milk the cows." She hesitated, then added, "You can join me if you wish," although her tone said she'd rather he didn't.

"That's all right. I need to talk to Seal, too."

"Our chief cook and gardener. You'll probably find him in the squash patch."

That was precisely where Drayco found Seal Hettrick, wearing a baseball cap with the motto "Not My Circus," and pulling up some multi-colored, gourd-looking thing. Drayco was three-for-three in the beard department with the commune males, although Seal's was scruffier.

Drayco introduced himself, and the other man said, "Gordon warned us you'd be coming."

Both he and Catherine had used that same word, "warned." What, if anything, should Drayco make of that? He picked up one of the squashes. "What is this?"

"Turban squash. Tastes like a cross between a sweet potato and hazelnut. You can cut off the tops and use them like bowls. I add them to soups and pasta."

Drayco put it back and grabbed a dark purplish vegetable. "And this?"

"That's a black beauty tomato. An heirloom variety. We try to grow heirloom seeds whenever we can. Decades of farm genetic engineering, and fruits and vegetables don't taste as good anymore." The man spat on the ground, making it evident the wad in his cheek was chewing tobacco.

"Where do you get the seeds?"

"Catalogs, mostly."

"I'll have to check them out. What were Lara's favorites?"

Seal scratched his chin. "Of my cooking, you mean? I'd say my borscht. And she also loved Smith Island cake."

"Did Lara tell you why she was leaving the commune? It was only three weeks before she was murdered."

"It's a jungle out there. We warned her not to do it."

"What was her stated reason for leaving?"

"Just wanted a taste of the party life, I suppose. She did like the finer things, though she tried to hide it. But I thought if that's what she wanted, it was oh-aard."

Drayco picked up on the unusual term for "okay." He said, "Oh-aard? Are you originally from Baltimore?"

Seal grinned. "Guess some things stick with you. Hard to escape certain bits of your past."

"People can have a lot of buried secrets in their past. Catherine said Lara had a talent for finding out such secrets. Did that include you?"

"Me?" He uttered a nervous chuckle. "Maybe my secret cigars from time to time." He spat out more of the tobacco with an apologetic look. "I switched to this when Gordon quit smoking. Seemed like the thing to do. Not tempt him with smoke blowing in his face all the time."

"What about your missing member, Ivon Leddon? Any secrets he tried to keep?"

"Hell, we still don't know for sure he's really missing. What if he wants to stay 'missing,' to forge a new identity?"

"Like commune members do by being here?"

"I suppose."

"Surely it's stressful for members to leave friends and family behind?"

Seal squinted up at him. "My wife died of a stroke at a young age. And my daughter. . .I haven't seen my daughter in a while. Sometimes, that's for the best. She has her own life."

"Gordon Aronson said members can travel when they choose. Apparently, several did."

"Oh, sure. Max is gone at least once a month since his parents are old and frail. But Lara, she never left much. Just to go into town."

Drayco reached down to feel a different type of squash, more gourd-ish. It had a firm, matte skin, with a few grayish, scaly lumps. "Members take something of a vow of poverty when they join up. Then where did Lara get the money to rent an apartment in Maryland?"

Seal dropped a couple more tomatoes in the basket. "Probably saved up. From trading and selling extra eggs and things in town. You know what they say about money stuffed under the mattress."

Drayco watched as the other man scooped up more of the turban squashes. They must grow like zucchini because a long line of plants waited to be harvested. "Seal's an unusual name."

"It's short for Sealey. Which I hated and got teased about. Until someone pointed out it sounds kinda like the Navy Seals, so it was cooler. And I went with it."

"How did Lara get along with the other members here?"

"Pretty well, as far as I know. I remember her staying up late playing cards with Max not too long ago. If she'd won, he was going to have to make a batch of her favorite pumpkin ale, which Max loathes. I think Max cheated just to get out of it. And as for me," Seal tossed a rotten squash aside and added with a smile, "Lara and I used to go bird-watching together at Powhatan Park. She loved it and could stay there for hours."

"And Lara's relationship with Catherine?"

Seal's smile dimmed. "Seemed okay enough."

Drayco pointed at the pile of squashes. "Since you're a good cook from Baltimore, I'll bet you make killer crab cakes."

Seal wrinkled his nose. "To be honest, I can't stand the things. I prefer coddies."

"Coddies?"

"Fried fishcakes made from cod, onions, and mashed potatoes."

With that bit of crab-cake blasphemy, Drayco left Seal to finish his task, grateful he wasn't the one picking vegetables in the hot September sun. Merely walking around the grounds had made him work up a sweat. Although he made a note to add turban squash to the list of new foods he needed to try.

Both Catherine and Seal seemed normal enough on the face of it. More like actual farmers than Wall Street escapees, and after being here for years, that was to be expected. But Drayco's years in the FBI had taught him to know when people weren't entirely truthful, and they were both hiding something, of that he was certain.

As Drayco headed back to the office to let Aronson know he was leaving, he passed by a small shed on the way. Curiosity getting the better of him, he peered inside. Looked like smoked fish hanging from hooks. More of Catherine's work? Or Seal's?

Once he rejoined Aronson, he had one more thing he wanted to ask the director. "I'm curious how you'd feel if a member squirreled away a bunch of money and hadn't taken that vow of poverty seriously?" He was partly referring to Lara, but also to Aronson, himself.

The other man frowned and looked down at his desk before starting up with his cuticle-picking behavior again. "We try not to be judgmental here. Besides, what would anyone do with money on a commune?"

Maybe Dennis Frischman was way off about Aronson being wealthy, but something about Aronson's evasive reply told Drayco maybe not so much. As he plodded back toward his car, he mulled over the commune members and their potential secrets—and wondered if Lara-Minna was as good at digging them up as Catherine had suggested.

But that also brought up another reason Harry himself could have killed Lara. . .because she had blackmail material on him. Doubly so since she vanished from Harry's life around fifteen years ago, which is about the time she went into the commune. Was she not as absent from Harry's life as he'd said?

Drayco opted for a detour on the way back to his car. He stopped by the brewery to talk to Max, who was hunched over another gauge as he worked on it with a wrench. When Drayco called out to him, he jumped. "You again, eh? More questions, or did you come for the beer?"

"Perhaps some other time for the beer." He took a whiff. "Although it smells pretty good in here. Actually, I was curious if Lara was blackmailing you."

Max stared at him. "You're rather direct, aren't you?" He dropped the wrench onto a nearby table where it landed with a loud clatter. "If you must know, Lara tried to nail me for money laundering. But she was misinformed. It was my old company, and it happened after I left. You can check it out, yourself."

"Thanks, I will." Drayco added, "At least coming to see you, I get my exercise."

"Legs or mouth?" Max said with a mocking smile.

"As Gordon put it, the brewery is a bit of a hike."

"Has to be set apart due to the volatility of the gases, you know. And because we were afraid local kids will come in and try to steal the beer."

"Is that the reason for the electric fence?"

"And all the graffiti and thefts we've had." He grimaced. "I was a teenager once, so I somewhat understand the lure of illegal booze and craving a little mischief. But at the risk of sounding like one of those 'get off my lawn' types, teenage pranks don't seem as harmless when you're on the receiving end."

When Max pulled a watch out of his pocket, Drayco took the hint and continued the trek back to his car. Now that he'd met the remaining four members of Gaufrid Farm, he mentally cataloged his impressions. One thing was sure. . .commune life must be good for one's health because none of them had a spare ounce of fat. Each one looked ridiculously fit, even Gordon Aronson, in his early sixties.

Drayco started up the car's engine and pointed his Starfire down the long gravel driveway, making mental notes of his visit as he headed toward U.S. 13. Engrossed in his thoughts, it took him a moment or two before the little tingling alarm in the back of his mind registered with full force. His senses on high alert, he spied something moving on the passenger-side floor of his car—or was it less like moving and more like slithering?

He veered over to the side of the road and managed to engage the brake before he flung open the door and vaulted onto the gravel. When he started to approach the open door for a better look, he jumped back as something lunged at him from inside. Instinctively, he grabbed a nearby rock and brought it down on the creature's head, which he now saw for certain was a snake.

The rock did the trick, and Drayco studied the tubular corpse lying on the ground. It had a reddish-tan-colored body with hourglass-shaped darker marks on its back. And a copper-colored head. Definitely not a friendly little green snake.

Were there more? He picked up a stick and carefully poked under the seats. After his pulse returned to something close to its normal rate and he was satisfied the car was now snake-free, he took stock of his situation. He was pretty sure there weren't any holes in the floorboard big enough to allow in a snake of that size. It could have climbed through the windows he'd left cracked at Gaufrid Farm to let in some cooler air. But were snakes known to be climbers like that?

If not, it meant someone might have tossed it inside his car. Each of the four commune members was alone at the farm at some point, separate from Drayco and with enough time to plant the snake.

Then again, maybe he simply needed to take the car to a mechanic when he got back and see if he'd missed some access point. He grabbed a small box and a towel from the trunk and carefully wrapped the snake inside. Even if not an intentional act, some might say it was an omen. Of what? To be wary of snakes in the grass?

15

Drayco sprinted up the stairs onto the porch of the white Victorian and opened the teal-and-glass door. He was surprised he wasn't greeted by cries of "Drayco sorry" from Andrew Jackson, the parrot, and the ordinarily immaculate interior seemed askew. The table in the foyer was pulled away from the wall with brochures scattered underneath. Letters on a display corkboard were arranged to read, "I donut get puns." And the floor was dustier than usual.

He sought out the Cape Unity Historical Society's director, amused to see the man had ditched golfer's plaid knickers for something akin to jodhpurs. "Reece, what happened here? Looks like a mini-tornado passed through."

"'Mini' is right! A tornado of tots."

"Tots?"

"School group. They were in this morning."

"Ah. Is that why Andrew Jackson is AWOL?"

"Andrew has picked up a few non-family-friendly words in his vocabulary. Plus, he doesn't like crowds."

"I hope he's not too traumatized by your titanic tot-nado."

"Alliteration R Us, Scott?" Reece scrunched up his forehead in faux outrage. Then he finally noticed Drayco was carrying something and pointed. "What's in the box?"

Drayco opened it and unwrapped the towel.

Reece cringed. "Dead, right?"

"Sorry to say. I acted on instinct. Otherwise, I would have let it go."

Reece put up a finger for Drayco to wait there and disappeared briefly, returning with a book. He flipped through it until he found what he was after. "Copperhead. One of the few venomous snakes on the Eastern Shore."

"I kinda figured. The coppery-colored head is a giveaway."

"You adding snake collecting to your hobbies?"

"I found it inside my car."

Reece shivered. "I don't even want one *outside* my car. How'd it get in there?"

"I'm not sure."

"Oh?" Reece tilted his head. "Maybe it had a little help?"

Drayco waved off the question. He'd tackle that later with help from his mechanic. "Did you find anything on Stuart Wissler?"

He wasn't exactly shocked when the notoriously thorough researcher answered in the affirmative. "Born in a small town in Delaware. The tiny local records office burned down years ago, alas. But I was able to dig up a few bits here and there."

"Here and there?"

"I have a colleague up in Maryland. He specializes in collecting historical gossip on all the notorious people up and down the Delmarva."

"Like our pirates from a few months ago?

"Not just pirates, but our own Bonnie and Clyde. Only they were named Doris and Fred. Which isn't as catchy and probably why you haven't heard of them."

"And Stuart Wissler?"

"Had organized crime connections, I gather. Or so my source surmised, nothing concrete. But as a minor player on the far, far fringes. More of a two-bit hustler who fancied himself a bigshot."

"He also liked to pass himself off as an attorney."

"Not surprised. Why are you so interested in the loser?"

Drayco went over to the board to rearrange the letters from "I donut get puns" to "stout gunned pi," and stood back to admire his work. "Wissler fooled at least one man into marrying a woman for green-card status."

"Let me guess. . .they skipped town with a bunch of said man's money."

"Fifty thousand." Drayco had nudged Benny to get the total amount from Harry. As if reacting to the cash stash, Drayco heard the parrot whistle loudly from a side room. He could swear that bird was smarter than it let on.

"Fifty grand isn't enough to live on. But if you do that several times—"

"You'd be pretty well off. Did your friend find any such frauds involving Wissler?"

"Not so far. I'll tell him to keep digging. He lives for finding salacious history stories."

"I don't suppose you found a current address for Wissler?"

"Somewhere out over the Atlantic, I think."

Drayco's jaw dropped. "Come again?"

"Died eight years ago from a heart attack. He was a heavy drinker and smoker and overweight. No signs of foul play. Didn't have any family, but on his deathbed asked a funeral home to cremate his remains and scatter them over the ocean."

Drayco's spirits sank at that news. A dead-end in more ways than one. "Do you have the name of that funeral home?"

Reece reached into his pocket and pulled out a piece of paper he handed over. "I thought you might ask me that."

"Thanks, as always. If you ever need my vote for Historian of the Year, just ask."

"We don't have awards. With all the dusty tomes, archives filled with spiders, and the occasional fossilized cracker, who could ask for anything more?"

Drayco pointed to a new painting hanging on the wall behind Reece, thinking he recognized the artist. "Is that a Virginia Harston creation?"

Reece beamed. "She's getting better all the time, isn't she? Pretty remarkable for a teenager."

"She's doing well? And Lucy?"

"Both daughter and mother are quite well, thanks."

From the way Reece's smile increased to full-on laser beam, it seemed Reece's relationship with Lucy was also doing quite well. Drayco asked, "You should have an art exhibit in here."

"Ah-ha, I'm way ahead of you. I've got one planned with Virginia's work, some of Barry Farland's, Antonio Skye's, and also Marty Penry's marine biology papers and artifacts. To showcase Cape Unity's talented young people."

"Should have guessed you'd be on top of it."

Reece gave a little bow. "Now it's my turn. How's the Opera House progressing? I haven't seen the contractor's crew lately."

"That's not as happy a topic. It's looking like the money I was counting on from the Chopin manuscript sale could be tied up indefinitely. If so, it won't arrive in time to save the Opera House project from bankruptcy. Or being sold."

"You know if you do that, a developer will bulldoze the whole thing down."

Drayco did know, and he was not at all happy at the thought. Sure, when he was first bequeathed the old building, he immediately thought *he'd* sell it, but he'd grown attached to the historic facility. His history and the building's history were now inextricably intertwined, and he felt like it had become an extension of his own soul.

He grabbed his snake box, thanked Reece again, and headed out down Route 13 toward a locally famous store. The Chesapeake Fish & Shell Company still had the same frame and weather-boarded architecture it did when established in 1920. Fortunately, this building had withstood attempts to tear it down, as well as several nor'easters and a hurricane or two.

When he popped inside, a young woman wearing a lobster-print apron greeted him with a big smile. "You're in luck. Our Chincoteague salt oysters are on sale today."

"I'll certainly have to check them out, then," and he dutifully headed over toward that section of the store. He picked up a box of the oysters and also looked at the price of pickled blue crab meat, Gordon Aronson's favorite. Those were sold in cases of six small jars for three hundred dollars. Too rich for Drayco's bottom line.

When he took his purchases up to the counter, he said to the clerk, "An acquaintance of mine, Gordon Aronson, highly recommends the pickled blue crab over there."

"Oh yes, he buys a lot of those, a case or two a month. Sends us a check, and we ship it to him."

"Do you need a business check to do that?"

"No, just a personal check."

When she looked askance at Drayco, he joked about Aronson's fear of banks to throw her off, which was only a little white lie. But the fact Aronson bought a "case or two a month" of pickled crab meat meant the guy was spending seven thousand per year just on these. Maybe Frischman was right about the commune director's secret financial stash.

Drayco toted his purchase to his car. Good thing they'd packed the oysters in ice because he had another visit he needed to make before heading back to the Lazy Crab. He was dreading it, but it had to be done. He reminded himself—like he'd told everyone else—he was a professional, and a client was just a client. Even if an ex-lover.

Before he climbed into the car, he gave it a thorough examination inside. No snakes or anything else that shouldn't be there. Until a gust of wind blew a leaf from a sugar maple through the open door and onto the driver's seat. As he picked it up to toss it aside, he saw it had a few tinges of red. The first leaf to fall in fall?

Catherine Cole would say it was "apportation," the spirit world manipulating physical objects here on Earth. Sloan DelRossi would spout some business-babble about using the leaf as an ideation to productize some company's personal brand. Hell, even Darcie might say it was a sign to buy herself a new autumn wardrobe. But sometimes, a leaf was just a leaf.

Out of curiosity, he pulled out his cellphone and got enough of a connection to do a quick search. According to folk tales, the first leaf of fall meant you should make a wish if you caught it. Did having it fall on your car seat count as catching it?

He rechecked the website info. It also meant good luck. Fine, then. He'd put that and his brick-rubbing from the Opera House into his good-luck bank. He had a feeling he'd need to make a lot of withdrawals. . .and soon.

16

The sun was half-mast heading toward twilight when Drayco drove up the long drive to Cypress Manor, its row of gleaming white columns blushing from the sunset's pink rays. When Darcie welcomed him inside, he noted the place still had the air of a partially inhabited museum with antique candlesticks on antique sideboards.

Although he did note Darcie had tamped down some of the more ostentatious displays. Plus, there were vases of fragrant roses and more feminine touches like red-and-purple silk sofa pillows and a rug embroidered with more roses. It was a bit unsettling for him to be there, with Darcie's ex-husband still in prison and her current fiancé in jail.

She greeted him with a hug that wasn't entirely sisterly but quickly released him with a small guilty smile. "Want some wine?"

"I'm driving."

"A wine spritzer, then. There's hardly enough alcohol to matter." She disappeared briefly to fetch one for each of them.

He looked around and noticed a few things he hadn't seen before. Boxes sat off in one corner, some taped and labeled with marker pens as "Dishes" or "Figurines." There seemed to be fewer tchotchkes, and her ex-husband's scrimshaw cabinet was also missing.

Drayco took a sip of the spritzer. Wasn't great, wasn't awful, but it was wet and cold. The temps had soared into the nineties, the last gasp of summer, despite the maple leaf. "Thanks for snagging the old photo of Lara Davidenko for me."

She swirled the liquid around in her glass. "I'm annoyed Harry kept it. Guess I shouldn't be."

"Since you're not in love with him, you mean?"

"And since I still carry the torch for you, darling."

"Surely he must check off some items on your husband-wish-list, or you wouldn't be marrying him."

"He's so charming and kind. And buys me anything I want."

"Isn't that what you said about Randolph Squier, your ex?"

"You think I'm shallow."

"That's kind of the definition of the word."

She sank onto a nearby red-plaid sofa. "For your information, I'm using my sociology degree to convince Harry to set up a foundation to help the homeless. He's going to give me the money to do it."

Drayco grabbed a chair as far away from her as possible. "Why the homeless?"

"I've got two homes now, don't I? Seems a shame some people don't even have one."

It was typical Darcie logic but still kinda nice, for her. He said, "If we can prove Harry's innocence, will you go on with the engagement? Knowing about this ex-wife of his that he didn't tell you about?"

"And I didn't tell him about you and me at first. It's the lover's code."

More Darcie logic, and again, he couldn't disagree. "Did Harry mention the name Stuart Wissler?"

She furrowed her brow. "I'm not great with names, but that's one you wouldn't forget. Rhymes with Kissler."

"Think back eleven days ago. The day of the murder. Did you see any unusual behavior from Harry?"

"He seemed perfectly normal. We were discussing wedding and honeymoon plans."

"He never hinted he had an appointment with Lara Davidenko that evening?"

"No. And I'm not sure why he didn't. You'd think that's the sort of thing you'd want to tell your fiancée, isn't it? That your ex-wife is coming over for a visit?"

"I would imagine so, yes. Unless it's the lover's code again."

She leaned back and kicked off her shoes. "You should stay here tonight at Cypress Manor." She stuck out her chin defiantly. "It's my place, and I can do whatever I want."

"I've got a room at the Lazy Crab."

"Oh." She bit her lip. "But the beds haven't been packed up yet. They're all cold and lonely." She winked at him, but it seemed half-hearted.

Perhaps she was beginning to "get over" him already. Or maybe Harry's situation brought out her protective instincts, making her feel guilty about a tryst with an ex-lover. But then again, she'd almost cheated on her first husband with Drayco, and months ago had suggested a three-way with Drayco and Harry. So there was that.

She planted her glass down on an end table. "Let me make you dinner, then. It's the least I can do for you. For taking on Harry's case."

"You can't cook."

"Neither can you."

"True."

She grinned. "But I might have picked up some takeout for two from the Island View Restaurant. It's warming in the oven."

Maybe that was presumptuous on her part, and this wasn't a good idea, but right then, he sensed neither of them wanted to be alone. She sat him down at the as-yet-unboxed dining table and carted out two servings of broiled stuffed flounder and coconut shrimp.

The flounder and shrimp were outstanding, even when reheated. It made Drayco realize how many frozen dinners he'd had recently, often not even having time for SiAm Thai Emporium takeout.

He asked for more details about her proposed charitable foundation and was surprised how thoroughly she'd thought the whole thing through. She'd spoken with accountants, and attorneys, and board members of similar groups. "I'm learning how to make spreadsheets," she said with a smile.

"Welcome to my world. I had to learn those, too."

"For your consulting business?"

"And the Opera House."

"Oh, yes, the Opera House. I keep forgetting about that." She studied his face. "If I've learned to read you at all, it's not going very well, is it?"

He shook his head, prompting her to add tentatively, "I could send some of the foundation money your way."

"No, absolutely not. The homeless population needs it more." He added, "But thanks."

He reached for her hand on the table and realized it still felt good. But it also felt different, somehow. She glanced down at their clasped hands, and when she looked up at his face again, he could tell she was thinking the same thing.

Not wanting to turn the moment into something too melodramatic, he said, "Spreadsheets. You have hidden talents, missy."

"Because knowing you has rubbed off on me. Just stepping into the room with you makes anybody's IQ jump fifty points."

"Hardly."

She eyed him pensively. "I wonder if anything from me has rubbed off on you?"

"You make me less serious. Being with you is like taking a mini-vacation."

"You have me there. You *are* too serious."

"And I'll never look at a copy of the *Kama Sutra* the same way again."

Her eyes opened wide, and she licked her lips. "Hmm. Unpacked bed. . ."

"I have a code, too. About not having sex with married or soon-to be married women. Or clients."

"But you and I got together before Randolph and I were divorced."

"Only *technically* before. You'd filed all the paperwork, and the divorce was essentially final."

"Spoilsport." But she was smiling at him.

"I confess, you did make me a 'spoiled' sport," and then he hastened to add. "But nothing, not even the universe, lasts forever."

She grabbed her glass again and downed the rest of the spritzer in one gulp. "Since you arrived, you haven't told me how Harry's case is going. Is it that bad?"

"I wouldn't say bad." But her question made *him* gulp down more of the spritzer. "I have a new lead on the victim. Harry's ex-wife may not have been Russian after all, but rather a con artist named Minna Hallow who's originally from Maryland."

"Seriously?" Darcie sat up straighter. "So the whole marriage thing was a scam?"

"It does look that way. But that's what I have to sort out. The who, the what, and definitely the why, or why then."

"Oh, my god. Does Harry know?"

"Not yet, but Benny Baskin and I have an appointment to see Harry in a couple of days. We'll discuss it then."

"My poor Harry. I can't believe anyone would take advantage of him that way."

"And you're sure Harry never mentioned the name of Stuart Wissler?"

She shook her head again.

"Gordon Aronson? Max McCaffin? Catherine Cole? Seal Hettrick? Ivon Leddon?"

"No, but who are those people?"

"Members of a commune Minna Hallow a.k.a. Lara Davidenko belonged to. I was curious if Harry had talked about any of them."

"Sounds like you've been busy." She studied him for a minute and then got up to give him a chaste—for her—kiss on the cheek before sitting back down again.

"What's that for?"

"For being you. And for being willing to take on Harry's case. Even though it must be hard. I mean, us," she waved her hand between them.

For once, having the subject brought up didn't make him irritated as it had when everyone else did the same thing. An existential relationship clearing-of-the-air? An acknowledgment that things really were over for good?

Or not so good, as the case may be. Because if Harry did turn out to be guilty—and Drayco wasn't about to voice those reservations to Darcie—then she might come running back to Drayco. Would he want that? And why did brief images of a certain deputy with blond, braided hair pop into his mind whenever he had those kinds of thoughts?

He grabbed his wine spritzer and let a little trickle down his throat, wishing it was something harder. He'd leave as soon as he could, not thinking at all about that bed, and get back to the safety of the Lazy Crab where the only temptation was Maida's Chickering piano.

17

Wednesday, September 23

Drayco looked over at the partially eaten gas-station breakfast sandwich on the passenger seat and wished he'd opted for one of Benny's S'mores doughnuts instead. Or maybe a breakfast burger with doughnuts for buns. Maida was always telling him he needed to fatten up.

After he'd checked out of the Lazy Crab, he'd opted to drive up Route 13 that snaked north from the Virginia end of the Delmarva Peninsula into Maryland, marveling at the flatness of it all. With few billboards or cellphone towers, the tallest objects were crepe myrtle trees on either side of the landscape.

The road didn't get near the Chesapeake or the Atlantic, so you could be forgiven for thinking you were driving through Kansas. That is, if not for the occasional whiff of salty air and the pungent rotten-egg smell from decaying algae in the marshes. It was a smell he'd sort of grown to like.

He reached over to turn on his cellphone in its holder and put it on speaker mode. Fortunately for him, Sarg didn't seem too busy to take his call. "Sounds like you're on the road, junior."

"Heading up to Brooksboro, Maryland. Population four thousand, one hundred and twenty-four."

"Beautiful downtown Brooksboro, eh? No, I'm kidding. Don't think I've heard of the place. Looking to buy a vacation home, are we?"

"It's the hometown of Minna Hallow."

"Ah. The fake Russian, fake wife, and real murder victim."

Drayco paused to exit onto Route 50. "*Какую запутанную сеть мы плетем.*"

Sarg replied, "*Ебать.*"

"Does your wife know you like to curse in Russian?"

"I haul out my rusty Russian now and then to do just that. So she won't know I'm cursing. I tell her I'm using Russian to stay in practice."

Drayco chuckled. "That makes more blackmail material for me, then. It's a pretty long list."

"Tell you what, I'll square it with you when I do whatever it is you've called to ask me to do."

"You wound me. Would I do that?"

"You already did, remember? About this same Minna Hallow?"

"And that's exactly why I'm calling. Did you find out anything about her Russian identity? Stolen, perhaps? Or about Stuart Wissler, for that matter?"

Sarg's tone turned apologetic. "As it turns out, Onweller dumped a big serial sexual assault case on me—the same day you asked me to check into Minna. But I haven't forgotten. I did make some initial calls."

"That's okay. I'll try to turn up something today that will make those unnecessary. Or narrow things down."

"Keep me posted. Gotta wonder, though. . .if the victim and Wissler crafted this green card scheme together, why not keep up that lucrative partnership? Why go into a commune for ex-Wall Street yahoos?"

"You got me. But Wissler *was* in on it because she used him as her 'character reference.' Blackmail's one possibility."

"Blackmail?"

"One of the other commune members hinted Lara was good at rooting out secrets."

"Ex-Wall Street members. . .that type would have a lot of those secrets."

"Insider trading, fraud, money laundering, regulation-busting of various stripes."

Sarg muffled his phone for a minute as he replied to someone who'd entered his office. The muffling over, he asked, "But she'd been there twelve years, right?"

"Fifteen."

"And the commune was founded twenty-five years ago. Seems like most of those secrets would have gone pretty stale."

"Begs the question of why these people would want to be part of the commune to begin with. Let alone for that long."

"Unless the commune isn't what it appears to be on the outside."

That matched almost word-for-word what Drayco thought, but they did say great minds think alike. At least, the minds of former FBI partners.

Drayco said, "There do appear to be plenty of secrets at the place. Then again, Wall Street is filled with NPD and APD types."

"Narcissism, sociopathy, psychopathy. I remember that research when it came out. You see any signs of that with these 'communers' of yours?"

Drayco considered that for a moment. Deceitfulness? Likely. Impulsivity, irresponsibility, and lack of remorse, the other symptoms of NPD? Possibly. He replied, "I'll have to get back to you on that."

"Good luck." A beeping started up on the other end, and Sarg said, "Gotta call coming in on the office phone. Just let me know what you find out, 'kay?"

Twenty minutes later, Drayco drove past a wooden sign with fading red letters spelling out "Brooksboro." As he drove through town, it reminded him of Cape Unity, but smaller and without the seaside charm. Where to start? The small food mart he passed? The place didn't have a Historical Society and probably not a historian par excellence like Reece Wable, but it did have a public library, according to the town's website.

He called up the library's address on his phone GPS and navigated to what looked like a converted white clapboard house with a green roof. A library sign hung underneath the porch roof, but otherwise, it felt like he was getting ready to step inside the home of someone's kindly grandmother.

When he went inside, the bookshelves lining the walls made it evident it wasn't a home, but the kindly grandmother part wasn't too far off. A woman with a silver braid curled around her head was filing books from a cart. As he got closer, he could tell her yellow dress was dotted with white owl figures.

He noted a nameplate on a small desk, "Ella Kerman, Librarian," and approached her. "Are you Ms. Kerman?"

She looked up at him. "You're a tall one, aren't you? Six-five?"

"Six-four. And a fraction."

"I loved fractions. Most girls I knew hated math, but I adored it." She held up a book, "This one's pretty good. Not math, but a natural history of wildlife in Maryland." She glanced down at her dress. "As you can see, I'm particularly fond of owls."

"I haven't come to check out a book. I'm hoping you could help me with another kind of history, one person in particular."

She set the book down and perched on the edge of the desk. "You've come to the right place if it's one of the townspeople. I've lived here my whole life, and I'm seventy-five."

"Was this building your home at one time?"

"Not mine, no. Belonged to a grandchild of the town founders. Then it was turned into a K-through-eight until they sent all the few remaining kids over to the county school. They weren't getting enough students to fill it up. I was a teacher in it for a time. English." She held out her hands. "Surprise!"

"Sounds like you might know pretty much everyone in town, then."

"Mostly the older ones. The newer arrivals, not so much. You said one particular person, though?"

"A young woman named Minna Hallow."

Ella nodded. "I remember Minna. Pretty little thing. Her grandparents brought her to this country when she was five. After her parents were killed during some USSR insanity." She tapped her foot on the floor. "Wait here a minute."

She headed to a set of wooden drawers filled with folders and flipped through them until she found what she was after. She carried her prize over to Drayco and handed it to him. "I have a file on most families in town. Not like a police file, mind you. Newspaper clippings, photos, articles, birth, wedding, death announcements. That sort of thing."

He opened the file and lying on top was a photo from the paper about Minna winning a high school beauty contest. Judging from the de-aged photo he'd made, this was the same girl as the one from the mugshot. Drayco carefully dug around through the folder but didn't see that particular item anywhere.

He said, "Hallow doesn't sound Russian."

"Oh, they changed it. Didn't want any prejudice following them."

"You don't recall their Russian surname?"

"'Fraid I don't. Minna was raised all-American. Cheerleader, baseball, cheeseburgers."

Drayco considered that image of the young Lara-Minna. All-American yes, but presumably the grandparents kept some of their Russian accent—enough for a young girl to be able to mimic. "She left town at some point?"

"She was around twenty when her grandparents died. Nothing much else keeping a girl like that around here, for sure. A bit star-struck. And a bit of a schemer, sad to say. Some called her a two-faced Tessie, but that's not what I saw. I saw a girl who was impressionable and naïve. Had big dreams like so many small-town girls do."

"Do you recall where she moved after she left Brooksboro?"

Ella pinched the bridge of her nose. "Not where, no. But with who, yes."

Drayco's pulse quickened at that. "A friend?"

"If you want to call him that. A fellow named Stuart Wissler, as I recall. That Wissler was a charmer in a sleazy, pomade sort of way. Lived here for less than a year, I think, doing odd jobs. And then he and Minna left. I lost track of the girl after that."

"Thanks for the information. I'm grateful for your time."

"All I have these days is time. Are you staying in town for a while?"

"I live in D.C. and drove up for the day."

"Too bad. We need some new blood in our gene pool. And from the looks of it, you've got some pretty good genes." With the way she was eying him, he didn't know whether to laugh. Or turn red. Or both.

Spying a photocopier, he asked Ella if it would be okay to make a copy of the news article about Minna Hallow to take with him, and she readily agreed. Copy in hand, he thanked her again and escaped to his car, thinking how Reece Wable would be salivating over all of Ella's historical documents. Maybe he'd have to connect the two of them. They'd get along famously.

So how did an "all-American" Eastern Shore girl come to be a con artist and commune member who winds up a murder victim? But Drayco now had a better idea of why she'd chosen a Russian persona. And a little more information to help Sarg in his quest for international intel.

ର ର ର

Bless Reece Wable's heart, the historian's research and directions were spot-on so far. Drayco stepped into Farley's Funeral Home and sought out the owner who turned out to be— unsurprisingly—a man named Bill Farley.

When Drayco introduced himself and asked about Stuart Wissler, Farley frowned. "That was an odd one, which is why I still remember it. Eight years ago, give or take. No family involved. Got a call from the guy while he was at the hospital. Said he didn't think he'd make it, and he wanted to prearrange his cremation."

"Did you talk to him in person or over the phone?"

"Had to go over there and make sure it was legit. The man handed over a wad of cash, enough for cremation. And some extra to pay me to dump the ashes at sea."

"Can you describe this man?"

Farley scratched the back of his neck. "As I said, it's been eight years. But it's not often I get a request like that. Burned in my memory, it is. He didn't look well, of course. Extra skin, like he'd been a bigger guy but lost a lot of weight real fast. Black hair and these big black glasses. You're not allowed to smoke in the hospital, so he kept chewing on this fat cigar."

Drayco compared Farley's description with his recollection of the grainy photo he'd found in the old news article. Sounded like it *could* be an older version of the same guy. "You say there was no family involved. Did you sense an estrangement of some kind?"

"Estrangement? He didn't say. I got more of an impression he just didn't have anybody. When I asked him about family members or friends who would wish to participate in the ash-scattering ceremony, he laughed. Said the only person he could think of who'd love to see him end up as dust was Hooty. And he wasn't about to give him the satisfaction."

"Hooty? No last name?"

"No, and I didn't push him." Farley gave an embarrassed chuckle. "We see an awful lot of interesting family dynamics in my line of work, Mr. Drayco. Not all of them pretty. I've learned when to press and when not to."

Drayco thanked him for his help and left with the knowledge that Reece's source about Wissler being dead and buried at sea was likely correct. But who was this Hooty? And did Wissler really not have any family whatsoever? Or perhaps he did and cut all ties years ago.

Kind of hard to maintain family ties when you're moving from one grift to another. Drayco had a sudden image of his own mother, and how Wissler's life and Maura's con-woman past sounded not all that different in some respects.

Wissler's molecules were floating somewhere in the depths of the Atlantic Ocean now. And Maura? Alisa's case had brought that topic back to the fore, even when he'd prefer not to think about where his mother was or if she was still alive. Not that it mattered—echoes of Drayco's mother haunted him in his dreams. Hopefully, with long days, lots of shoe-leather detective work, and some lucky breaks, he'd be able to keep Alisa from suffering the same fate.

18

It was after five when Drayco returned to his townhome, so he decided to make a snack. He wasn't really hungry, though, despite having consumed only the half-eaten breakfast sandwich earlier that morning, plus a bunch of coffee. He rummaged through his kitchen cabinets and cobbled something together that might do the trick. But he'd only taken one bite when his doorbell rang.

He was surprised to find his guest was none other than Nelia Tyler. She was dressed in a black short-sleeved sweater and black pants that made her look even thinner than usual, as if she'd dropped some weight. He was afraid this nightmare schedule of hers would eventually take its toll. He'd have to sic Maida on her, to fatten up Nelia, too.

She had a small briefcase she handed over. "A gift from Benny. Which means lots of papers and reading material."

"Thanks for bringing it by. I think. Let me get you a drink for your troubles."

"If you have a soda, that would be welcome."

She followed him into the kitchen where he grabbed a Manhattan Special soda from the fridge. She eyed his snack on the counter. "What, no fluffernutter?"

"Ran out of ingredients."

She walked over to get a closer look. "Ugh. Is that hot dogs in applesauce?"

"Guess I haven't had time to make it to the store lately."

"You do know they have something called carry-out now? Why, they'll even deliver right to your door. It's amazing. You should try it."

"You've been exposed to Benny too long. His sarcasm is rubbing off."

She looked around the place. "Are you by yourself?"

"Me and the piano, why?"

She rubbed her arm and uttered an awkward laugh. "I was half-expecting Darcie to be here."

He shook his head, more from exasperation than anything. "No, but I visited Darcie at Cypress Manor yesterday while I was on the Eastern Shore. She wanted me to stay with her there, but I didn't think it was a good idea."

"Hmm." Nelia gave him a skeptical look.

"We did have dinner."

"And?"

"And it felt. . .odd, different."

"Good different or bad different?"

"Just different."

"It's nice you're not at each other's throats. Breakups can be tricky." A shadow passed over her face.

Drayco hated to see her like that, with a raw vulnerability. He fought the urge to wrap her in a hug, but that was a really bad idea for a whole host of reasons. "How are your parents doing?"

"Good, they're good. Better than I am. I gather they'd been expecting their divorce for years. And I was so blind, I never saw it."

"It'll get easier in time."

She sighed. "So they say."

Drayco gestured toward the den. "Rest your weary bones for a few minutes. It's not good for you to gulp down one of these," he said as he picked up another espresso soda.

She plopped down on the sofa. "Benny told me you'd called to tell him Stuart Wissler was deceased."

"Reece found that out, and I verified it with the funeral home. Eight years ago, heart attack." He sat in his favorite red chair opposite her.

"Still, I have to wonder if the guy's as dead as we think? I mean, what if he faked his own death? Maybe there's someone else in his coffin?"

"His urn, actually. According to the funeral home, he was cremated and his remains allegedly scattered over the Atlantic. He'd left instructions and pre-paid. No family was involved, and the funeral home owner was okay with it. Probably got a nice day of fishing out of it at the same time."

That made her tense shoulders relax, even if she didn't crack a smile. "Eight years ago and no exhumation. Means my faked-death hypothesis is still in play."

"True." He grabbed the nearby ottoman and put his feet up. "Did Benny tell you about Harry's daughter?"

"He did, and I think it's very odd. Why now?"

"That makes two of us. But she did seem to be honest about that whole genealogy thing. I've dug into her background to verify her family details—the adoption, her father's death, her mother moving to Sweden."

"She's a biology grad student like she claimed?"

"Solid A-student. But on the other hand, she's also a struggling actress in addition to being a student. Been in some school drama productions and local community theater."

Nelia used her fist to pat her chest. "You men are lucky you can belch out in the open."

"You can, too." He waved around the room. "It's just you and me. No one to judge."

She finally grinned. "You find belching women sexy?"

"Scratching and spitting, too." He grinned back.

They sat in companionable silence for a moment until she asked, "Okay, what if Alisa Saber decided to come forward now to get some of Harry's money as his sole heir? Especially since Harry's ex-wife is conveniently deceased, and he's not yet married to Darcie."

"I did look up Alisa's current address. She got a full scholarship to Georgetown, so that's how her tuition was paid. But she lives in a pretty humble apartment. So, it's a possibility."

"Any other motives you've thought of?"

"Perhaps she wanted to frame Harry for abandoning her biological mother. And killed Lara because she took the place of Alisa's biological mother, in a sense."

"If that's the case, sounds like she's majoring in psychopathy, not biology."

"There's a lot of that going around. At least, the psychopathy part. Anytime you deal with Wall Street figures."

"The commune?"

"Everyone there seems to be hiding something. I don't know whether it's relevant to the murder. But again—"

"Timing."

"The victim moved out, three weeks later, she's dead."

Nelia took another swig of the soda. "But it would be impossible for any of them to drive over that far, do the deed, and then get back without anyone knowing. Even if they were all confined to cabins for that alleged virus."

"I can't rule out former members, either. Ivon Leddon, for instance."

"The missing man?"

"Thanks to Sarg's assistance, there don't appear to be any foreign passports he may have held, legal ones, anyway. And no attempts to use his legal passport since his disappearance."

"So if he did skip the country, it was by illegal means. There are several ways to do that if you're so inclined."

"Boats, private jets, border crossings." Drayco patted his chest and let out a belch. "See? It's easy. Are you going to join Benny and me tomorrow morning to visit Harry in jail?"

"I've got some studying to do for a test. Benny said he didn't need me. But it's why I had to deliver those," she pointed at the briefcase she'd brought, which was lying on a table. "It's more of the Fairfax PD's case against Harry."

"Which, on the surface, seems fairly damning."

"Benny's still optimistic."

"Because he doesn't want to admit his perfect defense record might be in jeopardy?"

"Because he believes in Harry's innocence. But more because he believes in you."

Drayco joked, "Guess I'll have to give Benny a nicer Christmas present this year."

All joking aside, he had to admit he was touched by that. Sure, the fact Benny used Drayco on cases at all should speak volumes. Benny was a no-nonsense person, not the type to put up with fools. He'd been known to drive a few interns, clerks, and paralegals to tears by his exacting standards and OCD procedures. Even Benny's wife once called him an abominable ass.

When Drayco noticed Nelia's now-empty bottle, he asked, "Want another?"

"No, I've got to get going. My studies and all."

"One for the road then."

She accepted the offer gratefully, and he waved to her as she headed out the door, soda in hand, and back to her waiting car.

He was a little bemused—and still a little exasperated—she'd expected to find Darcie here. Especially knowing Darcie was engaged, albeit to a man currently behind bars. Drayco was no saint, but he wasn't about to take advantage of a vulnerable woman like that. Surely Nelia would have expected no less of him?

With a heavy sigh, he grabbed his briefcase and pulled out the photo of Lara Davidenko from Harry during their "marriage." Then he grabbed the de-aged photo of Lara, as well as the newspaper clipping from Ella Kerman, the librarian. Absent a DNA test, it would be hard to verify beyond a shadow of a doubt, but he was still convinced Lara and Minna Hallow were one and the same. Ella's recounting of Stuart Wissler's connection to Minna sealed the deal.

A fake-Russian con woman, a rich and naïve businessman, a commune filled with characters allegedly escaping from Wall Street excesses. The only common thread was Lara-Minna, and she wasn't talking. But then, that's what Drayco was paid for, was it not? To get the dead to talk, to tell him their hopes, their fears, their failings. . .and to give up their darkest secrets.

Feeling he was diving too deep in philosophizing, he got up to eat more of his "dinner," but then chucked it into the trash. Nelia was right, it was pretty horrible, only slightly better than prison food. Wonder what Harry Dickerman was having for supper tonight?

Drayco picked up the phone to call SiAm Thai Emporium for some delivery and resigned himself to an evening of frustrating and very dry reading, thanks to one pint-sized attorney. Maybe Drayco would put some cayenne pepper in Baskin's S'mores doughnut next time. Then again, knowing Benny, the man would probably like it.

19

Thursday, September 24

While Drayco waited for Benny Baskin to arrive at the Fairfax detention center—no doubt picking up some breakfast coffee and S'mores doughnuts from Hava Java, minus the cayenne—Drayco studied the blue chairs and white tables in the lobby. The decor was on the plain side but still fancier than Sheriff Sailor's digs in Cape Unity. Of course, the fact Fairfax County PD had an annual budget of a hundred eighty million while Prince of Wales County's was more like thirteen million might have something to do with that.

A few other people occupied lobby chairs, but none seemed to pay much attention to him or the surroundings. He studied his fellow waitees. The forty-something woman sitting ramrod-straight while occasionally dabbing at her eyes with a tissue was likely here to see a prisoner—judging from the lack of a ring, possibly a boyfriend or adult son. A youngish man to Drayco's left wore a slightly wrinkled suit with a basic black briefcase at his feet. A public defender?

After the hands on the wall clock had ticked away fifteen more minutes, a female guard came to escort the teary woman through a metal detector. Shortly after, another guard lumbered in and called out, "Mr. Keene?" And when the wrinkled-suit man stood up, the guard added, "Time to meet your new client." Guess Drayco was two for two.

A man coughing to his right got his attention, and he looked over at Detective Shephard King as he crossed through the lobby. The lawman spied Drayco and charged over. Drayco steeled himself for a diatribe about the victim's trashed apartment, knowing the FPD was none too pleased it was Drayco and Sarg who were on the scene first.

But King surprised him. "Thanks for the Gaufrid Farm tip. But after me and my colleagues questioned the staff, doesn't seem to be any way they could be involved. Too far away, and the timeline won't work."

"There's the missing man, Ivon Leddon."

"We dusted the crime scene. No prints from any strangers. So that's a no to Leddon. Also, no neighbor sightings of anyone unusual, before or after."

"As you noted with Darcie Squire, Leddon could have hired someone to do it. Or he simply used gloves. Plus, there's the back fence and the side door."

"You got a bunch of 'ifs' and 'maybes' there that feel too much like forcing square pegs through round holes. We have our round peg in custody, I'm pretty sure of that."

King acted like he was going to add something. But when he took note of a legal dynamo headed toward them, he scrambled away so fast, he almost left a vapor trail behind. Said legal dynamo apparently put the fear of God into police detectives, too.

Benny Baskin stopped in front of Drayco and looked in the direction of the departing King. "Giving you grief, is he?"

"Not too much." Drayco pointed at the corner of Benny's mouth. "I think you've got a little crumb of sugar there."

Benny used his sleeve to wipe it off. "You let me know if these FPD guys lay into you. I can keep them at bay."

"Nice to know I have an attack-Benny at the ready."

They went through the usual check-in procedures and were eventually led to the same consultation room as before. Just like before, Harry was soon ushered in and seated at the table with them.

Harry said right away, "Please tell me you've got good news. Sitting around all day in a tiny windowless cubicle and not knowing what's happening out there is driving me nuts."

Benny exchanged a glance with Drayco. "We do have news. Whether it's good or not remains to be seen. It appears your fake ex-wife was also a fake Russian. Her real name was Minna Hallow, and she was raised in a small Maryland town."

Harry's eyes widened. "But her accent was so realistic."

"She was brought here at the age of five by her grandparents from Russia. Easy to parrot an accent if you'd heard it while you were growing up."

"But why pretend to be Russian?"

"To play on your pity, perhaps. More of that damsel-in-distress thing."

Harry looked down at his manacled hands on the table. "They pushed the right buttons. I was primed to fall for a Russian woman."

Benny opened his mouth as if to ask for more details, but Harry hastened to explain, "I told you about that party where I met a woman from Russia. God, it's been twenty-five years now.

Benny asked, "No name at all?"

"Tasha something, she never told me her last name. It was at Congressman Curt Goldfeder's house, no less. We danced for hours, she and I, as if no one else was in the room. Long story short, I had sex with her in a bedroom upstairs. Sounds sordid, I know. But it really was love at first sight. I'd never believed in it before."

"What happened afterward?"

Harry looked up at the ceiling. "I fell asleep, I woke up, and she was gone. Would have searched to the ends of the Earth, though I'd only known her a short time. But it was as if she was a ghost. Anyway, that's probably why I was so susceptible to Lara's story ten years later. And why I was taken in by the whole fake-Russian act."

Drayco exchanged another glance with Benny and then said, "It's interesting you should mention all this. Because a young woman named Alisa Saber walked into my office a few days ago. She claims to be your daughter. And she's twenty-five years old."

Harry sat there, stunned. "She could be lying."

"She did an ancestry test. Tied her to a cousin, your brother Mike's niece."

"Then maybe it's Mike's daughter."

"Your brother was stationed in Okinawa around the time Alisa would have been conceived. And there's more. The test showed another cousin brought from Russia by an adoptive family, meaning a matrilineal Russian link. Alisa tracked a worker at the adoption agency where she was dropped off as a baby. That worker said the woman who handed her over spoke with a thick accent."

Harry blinked slowly. "But why come forward now? Is this girl angry with me for abandoning her mother? Wanted to frame me for her murder?"

"She's a biology student, for one, so genetics is a passion of hers. After she got the ancestry results, she read about your arrest in the paper and came to see me. She wants me to prove you're her father. And to find her mother."

"Her mother?" Harry frowned. "You think it's Tasha?"

"I didn't know about your relationship with Tasha before now, but it fits the daughter scenario in a circumstantial way."

"I have a daughter." As he sat there fixated on the table, Harry's eyes glistened with tears for a moment. But his expression soon turned darker. "I don't want to see her."

Drayco stared at him. "Why not?"

"I don't want her to see me this way."

"Again, why not?"

"Because my own father was convicted of a crime when I was sixteen. I still remember the shame and anger when I visited him in jail."

Drayco replied slowly. "We can't be one hundred percent certain she's your daughter without a paternity test."

Harry slouched down in his chair and sighed. "I don't have a stellar record with the fair sex." He motioned toward Drayco with his hands, as much as the cuffs would allow. "I know about you and Darcie. That you were lovers. That's why I wasn't thrilled to have you on this case."

"It wasn't serious, and it's over."

Harry grimaced. "Not that I'd blame her. You're young, she's young. It gets harder to satisfy a woman as you get older."

Drayco blurted out without thinking, "Please tell me you're not suggesting a three-way."

Benny looked shocked for once, but Harry had a small smile on his face. "No, but I'll bet she mentioned it to you already. I have no delusions about Darcie being in love with me. But in time, I hope she'll come around."

Drayco was more than eager to change the subject. "When you were together with Lara, did she mention a man called Ivon Leddon?"

Harry looked confused and shook his head.

"What about Gordon Aronson, Catherine Cole, Max McCaffin, or Seal Hettrick?"

Again, Harry shook his head. "I don't know any of those people. Unless it's under some other name since Lara's was fake."

"And you never heard a place called Gaufrid Farm mentioned at any time?"

"A farm?" That made Harry chuckle briefly. "The most I know about farming are the fruits and vegetables that get delivered to my house from a farmer's market. Why do you ask?"

"Gaufrid Farm is the name of a commune Lara joined right after she stole your money and disappeared. She used Stuart Wissler as her reference to get in there."

"That shyster? Oh, man, was I played. And hard. But why would a couple of con artists join a commune?"

"Wissler didn't join. Only Lara."

Harry clenched his jaw. "But that makes even less sense."

Drayco couldn't disagree there. "We're working on it. And other angles, rest assured."

When the guard came to escort Harry back to his cell, Harry added, "Don't forget what I said. I don't want to see this Alisa person. Whether she's my daughter or not."

Drayco and Benny made their way to the lobby where Shephard King was nowhere in evidence this time. Probably cowering under his desk. The baby.

Benny thumped Drayco on the back. "Well, boy-o, find that real Russian woman of Harry's if you can, since she may have ties to this case. And maybe you'll find Alisa's mother in the process, win-win."

"Harry might be right. I've considered the notion that Alisa's timing isn't coincidental. Poor Harry seems to attract conniving women."

"Sure, sure. Pieces of a puzzle, yada yada." Benny gave Drayco a little side-eye. "Umm. . .a three-way?"

"Trust me, you don't want to know."

"Call me old-fashioned, but in my day, people didn't talk about sex. Even a lady's bare ankle was verboten."

"You are not that old, Benny."

"Okay, my grandfather's day, then." Benny cackled. "Lailani's gonna love this."

"Don't you dare mention that three-way business to your wife."

"I could be tempted to forget all about it. With the right bribe."

Drayco grinned. "Two boxes of S'more's doughnuts?"

"Throw in a dozen Krispy Kremes, and you've got yourself a deal."

20

Drayco pulled into a parking lot that seemed to lead to nowhere. That was fitting since he'd driven around what felt like nowhere for a half-hour. He finally found the small unincorporated area called Peaceville on the fringes of Prince William Forest Park. It wasn't all that far from Quantico, but despite his years at the Bureau, this place hadn't registered on his radar.

He started down a walkway of sorts, surrounded on both sides by thick vegetation that threatened to pull at his shirt and hair. No signs nearby identified the property, but he was certain this was the right place. The walkway turned out not to be as long or lead into nowhere as he'd first calculated, and he finally saw a small wooden sign to the left of the walkway that read "Monastery of the Peace Brethren."

Drayco headed for the first building, which looked office-like, a very basic, very square wooden structure painted white. 'Tis a gift to be simple and all. When he announced to the brown-robed man inside that he had an appointment, the fellow escorted him to a more modern-looking structure, an A-frame with a red tin roof.

The monk who'd guided him bowed without ever introducing himself and headed back along the path they'd come. Drayco didn't know whether to knock on the A-frame door or not, so he tried the knob, and the door opened.

What it revealed wasn't totally unexpected, given what Max McCaffin had said. The man in the corner of the room didn't have on a brown robe, instead sporting black slacks, a white smock, and a hairnet despite not having much in the way of hair to net. And he was stirring a yellow-whitish liquid in a large silver kettle.

He looked up when Drayco approached. "You must be Scott Drayco. That is, I assume you are since we don't get many visitors."

"And you're Daven Monk." Drayco paused, then added, "I know you get this a lot, but. . ."

"Yes, it's my real name." He chuckled. "Destined for the calling."

"But you were most recently at Gaufrid Farm. And before that, a Wall Street trading company called MacDuff Partners."

"All true."

Drayco eyed the vat of creamy liquid Monk was stirring. "Cheese?"

"Gouda. The best kind."

"Do you prefer cheesemaking to beer brewing, like you did at Gaufrid?"

"Oh, I miss the beer, all right. Making cheese isn't all that bad, it's just a different aging process." He excused himself for a moment to stir the liquid some more, then he turned back to Drayco. "After all, blessed are the cheesemakers."

Drayco smiled. "*Monty Python and the Holy Grail*?"

"One of the best comedy movies of all time."

Drayco studied the man's face. "No beard. After seeing Gordon, Max, and Seal with their facial hair, I could be forgiven for thinking Gaufrid was Amish country. Did leaving the farm mean leaving your hair behind, too?"

Monk laughed "Had Hippocratic baldness since my thirties. But yes, I did have a beard I shaved off when I entered the monastery. Symbolic? Who knows?"

"I'm curious why you left the commune. Two years ago, I believe."

"Two years next week. As to the why, well, have you seen the place?"

"Recently, yes."

"It's steadily fallen into decline. Not that Gordon doesn't try his best. But I think its time has passed. And this," he waved his hand around, "is a higher calling. Not the cheese so much."

Drayco studied the kettle for a moment. The liquid resembled a thick yellow chicken soup but smelled like dirty gym socks. The Gaufrid beer perfume was far better.

Monk had also been studying Drayco and blurted out, "I know why you're here. It's a long time coming."

"You knew about the victim's murder in advance?"

Monk's eyes widened. "Murder? You said this was about Lara, but you didn't say anything about murder."

"You hadn't heard about that?"

"No one called me. And I wouldn't have seen the news otherwise since I avoid that for the most part." He crossed himself and said softly, "'In this world you will have trouble. But take heart. I have overcome the world.'" He added, "From the book of John, chapter sixteen."

"Then you assumed I came to ask you about blackmail, am I right?"

Monk hesitated for a fraction of a second. "Cocaine. Using and dealing. I gave it up years ago, but when Lara found out, she demanded cash. Or she'd tell Gordon, then the police, and my family."

"How much cash?"

"The first few years, it was a few hundred here and there. All told about seven grand. I'd had some savings I brought with me that had dwindled down. That was the best I could do."

"Director Aronson didn't find out?"

"He didn't have to, I finally told him. And I told Lara I didn't care who she blabbed to, I wasn't going to pay her any more funds."

"What about the monastery? And any vows you had to take here?"

"I had to tell the order the truth. As part of confessing sins prior to being accepted."

"Do you know of others at Gaufrid Farm who Lara was blackmailing?"

"That would fit her personality. Always scheming something. Lack of self-confidence, if I want to put on my amateur psychologist's hat."

"Catherine said Lara had a talent for digging up people's secrets."

Monk snorted derisively. "Yeah. But if she treated others like me. . ." He rubbed his forehead. "I gather she'd never read Matthew chapter six, 'Don't store up your treasures here on Earth but in heaven.'"

"Since you're in confession mode, let's talk about Niles Peto, another commune member. I'm told he died of a cocaine overdose. Any connection to your drug use?"

"Niles and I did some coke together, sure. But it was recreational. One of the reasons I quit was when he died. We both kept saying we had a handle on it. That we were safe."

Drayco picked up a cheese wheel perched on top of a rack of wheels, row upon row. It did look pretty good. "Ivon Leddon went missing eight months ago. Were you also unaware of that?"

"Gordon called me after Ivon vanished. To see if he'd contacted me."

"Were you and Ivon close, then?"

"He's a great guy. Open, honest, helpful. Not depressive, if anyone is thinking this is suicide. He liked to talk about his family in Travilah. His parents are gone, and he didn't have any siblings. Did have some cousins, though. Porgy was his favorite. Named after the opera because his mother was a fan."

"Was he an honest enough fellow that Lara wouldn't have blackmail fodder on him?"

"Would he have told me if he did? Who knows?" Monk leaned against a tall cabinet. "You haven't asked me about an alibi for whatever time Lara was killed. That's why you're here, isn't it?"

"Have you left the monastery in the last two weeks at all?"

"Not once."

"Any witness corroboration for that?"

"We don't have cars here, not even like the sole communal car Gaufrid uses. What day was this?"

Drayco told him, and he replied, "Fridays are set aside for a day of meditation."

"Communal or private?"

"Private in our quarters or in the woods, wherever we find it most meaningful." He frowned. "So her killer is still at large?"

Drayco said, "The police have arrested someone."

"Obviously, you don't think they have the right guy, or you wouldn't be here. Who is it?"

"A man she once conned into marrying her for a green card and then disappeared two weeks later with some of his money."

"That must have been before she joined Gaufrid. Which was. . ." Monk stopped to do some calculations. "More than ten years. Twelve, maybe."

"Fifteen."

"Time seems to stand still when you're in a place like that. Or this. No work commute, no weekend binges, no appointments, no must-see TV. 'Therefore do not worry about tomorrow, for tomorrow will worry about itself.' That's from the book of Matthew."

"What did you know about Lara and her past?"

"Very little. Just that she was fleeing some fellow in a Wall Street firm who'd abused her." Monk peered at Drayco. "But since she conned that guy into marriage, I'm thinking that was another lie."

"Turns out Lara Davidenko's real name is Minna Hallow, and she grew up in a small town in Maryland. As far I can tell, no Wall Street connections whatsoever."

Monk rubbed the back of his neck. "Her accent was pretty good."

"Her grandparents were Russian. Her parents, too, but they died when Lara was young."

"Wonder if Gordon knew? He was besotted with her. I always thought it was because of that sexy accent."

"Besotted like a crush or something deeper?"

"I'm not much of a judge of relationships. Mostly. It was far easier to tell that Catherine was in love with Gordon."

Drayco made a note of that. He'd already had such suspicions from the way Catherine had talked about Aronson. "If Catherine was in love with Gordon—who was in love with the victim—were there signs of jealousy?"

Monk glided over to the cheese kettle to give it another stir. "Some, but Catherine kept it low-key."

"Why the director? There were other eligible men around for Catherine to fixate on. You, Seal, Max."

"Probably the father figure or messiah figure thing. Students falling for their professors."

"Did any of the men harbor romantic interests toward either of these women? Or the director?"

"Love? No, I don't think so. And we didn't take a vow of chastity to join up, either. Plenty of sex to go around. We all played 'musical cabins.'"

"Catherine hinted at that."

"Any time you wanted it, it was there. No attachments, just sex."

"Catherine hinted at that, too." Drayco edged closer to the cheese vat to get a better look. Cheese sludge was more like it.

"Why do you think I left that place, Mr. Drayco? Those vows meant squat. Any time a member left to go to town or to visit family, they could do whatever with whomever. And since we had an open-door policy, I saw a lot."

"What about other secrets?"

"I had a don't ask, don't tell philosophy. Here, everyone knows my secrets. It's a relief to unburden yourself from all that baggage. True repentance and reflection."

Monk handed the paddle to Drayco to try his hand at stirring. It was thicker than it looked.

Drayco said, "Gordon Aronson said something similar about Gaufrid. That it's named after a German phrase meaning the land of friendship, the land of peace."

Monk went over to rearrange some cheese wedges, rotating them on a rack. "Gaufrid turned out to be more cesspool than peace—blackmail, jealousy, squabbling. You asked why I left, there you go." He hastened to add, "Not that I'm judging. 'Why worry about a speck in your friend's eye when you have a log in your own?' That one's from the Gospel of Luke."

"You said Gordon was 'besotted' with Lara. Enough to be angry with her for rebuffing his affections?"

"And killed her when she left?" Monk shook his head. "I don't believe that. Although. . .I'd heard them arguing more before I moved here. Lara and Catherine, too. I know how that sounds, all that catfight stuff."

"Did Lara argue with Seal and Max, too?"

"Don't recall it. She and Seal were more like brother and sister in a way. And she and Max. . ." he laughed. "Two peas in a pod. They liked to see who could drink whom under the table."

Drayco handed the paddle back and pulled out one of his business cards. "If you think of anything else that might be helpful. . .you do have access to phones here, right?"

"We do." Monk took the card. "And I'll pray for Lara and the man they arrested."

"The real culprit, too?"

"There are enough prayers to go around, Mr. Drayco."

Drayco left the way he'd come after Monk told him he wouldn't need an escort. He had more questions than before he'd arrived, but the blackmail angle was looming larger on the motive list. And the missing man, Ivon—did Lara murder him but was later killed herself by an accomplice to ensure her silence? Maybe Ivon Leddon had a little blackmail on Lara. . .blackmail that went both ways.

On the other hand, Drayco couldn't rule out the possibility Gordon Aronson found out about Lara's blackmailing and—besotted or not—decided to get rid of her to keep the commune from dissolving. Or did Lara have something on him she recently confronted him with? Murder, perhaps? After all, the book found beside Lara's body had a torn page on the subject of murder and accessory.

When Drayco reached his car, he pulled out a copy of the photo he'd made for himself of Minna as a young woman. Despite it being a mugshot, the eyes had a defiant sparkle to them, but also a hint of impishness as if to show to the camera, "It's all just a bit of fun we're having, wouldn't you say?"

Star-struck Minna Hallow from a small Maryland town had definitely not found fame and fortune but instead slipped into a life of petty crime, from marriage scams to blackmail. Sadly for her, she got sucked into a cesspool of criminally toxic waste. . .and fell in way over her head. With that depressing thought, he tossed the photo back onto the passenger seat.

Spotting some of Major Jepson's coastal sweet pepperbush nearby, he snapped off a small sprig. It was also called "poor man's soap" because folks once rubbed the flower spikes together for a soap substitute. Maybe it was a fitting plant to grow beside a monastery where the point was to wash away your sins through work and prayer. He took a whiff of the spiky white flower. It really did live up to its name with a tangy sweetness that was quite appealing.

He tossed the sprig next to Lara's photo and plotted his next move. It would take a bit of a drive, but he just hoped it paid off.

21

After digging into the same databases of old news articles where Drayco found the grainy photo of "Stu" Wissler, he'd uncovered a lead on "Hooty," the friend of Wissler's that Funeral Director, Bill Farley, had mentioned. It took a lot more digging until he'd found a possible address in the unincorporated town of Massaponax.

But the building Drayco pulled up to didn't inspire confidence, nor did the bearded guys smoking joints sitting on Harleys in front of the place. He was pretty sure he had the right location, despite the lack of identifying signs. The site could be a brothel, a bar, or a bank, albeit a shady one.

He slid out of the driver's seat and studied the bikers who were studying him in return. The man nearest him, sporting a tattoo on his neck of a red-eyed laughing demon, pointed at Drayco's Starfire. "Nice car."

"Thanks. Runs pretty well."

"A '63?"

"1962. The engine is much younger, but it needs some work." It always needed some work. The never-ending repair bill.

"We don't have mechanics here."

"Actually, I'm looking for a man named Mark Nascha Owling."

Demon-tattoo-man squinted at Drayco. "Might know somebody by that name, might not." The man gave an almost-imperceptible eye shift toward the building behind them.

"Tell him I've come to swap old Stu Wissler stories."

"Wissler?" The man's face didn't register any recognition whatsoever.

"Mark Owling knows him."

Tattoo-man grunted. "Gimme a minute." And then he disappeared inside.

Drayco leaned against the Starfire, marveling at the building's "architecture," if you could call it that. Shaped like two large upside-down measuring cups back-to-back, or what would happen if a pyramid and a rotunda mated. Looked like it had been around for decades without much updating, unlike the Starfire.

After three tense minutes with more of the staring contest between the bikers and Drayco, Demon-tattoo-man returned to say, "Father Owling said he'd see you."

Father Owling? That's when Drayco spotted a small, cross-shaped window cling in the corner of one pane. He followed his guide through a series of hallways, surprised to see how much bigger the place was than it appeared outside. When they reached their destination, Drayco rubbed his eyes to make sure he wasn't dreaming.

The room felt like he'd transported back into the 1960s to a few years after Drayco's Starfire was "born." It was a tie-dyed paradise of purples and greens and reds, with tangerine-colored bead curtains and yellow bean bag chairs. As acoustic guitar music wafted through the air, he spied a turntable in a back corner, and the aroma of pot and sandalwood incense made his nostrils burn.

A man with long, braided silver hair sat in the lotus position on one of the bean bag chairs, his palms upturned, and his eyes closed. After the guide left Drayco alone with this man, he opened his eyes, clouded with cataracts.

Drayco asked, "Are you Hooty Owling?"

The man lifted one eyebrow a fraction of an inch. "No one's called me that for years. Not since Stu. But then, Reginald did tell me you said you'd come to swap old Stu Wissler stories."

Drayco tried to reconcile the name "Reginald" with his tattooed biker-guide, itching to find out more about the guy's origins. He replied to Owling, "Why did Wissler call you Hooty?"

"Isn't that obvious? Because of my last name. Well, that and my middle name, too."

"Nascha? Sounds Russian."

"It's Navajo. My parents were into Native American spirituality. Although they got it wrong. Nascha is the feminine form of the word meaning 'owl.' Nashota is male. Owl. Owling. Hooty owl. Stu thought it was hilarious."

"Reginald called you Father Owling. Is that an honorific?"

"I really am a seminary-trained, ordained priest. The motorcycle gangs who come through this area need a little spiritual nurturing now and then. Plus a place to bunk. And I don't judge."

Drayco looked around for something normal to sit on, but he noted only bean bag chairs. Resigned, he lowered himself onto a bean bag across from Owling. It was only slightly less uncomfortable than it looked and definitely not designed for long legs. Cold and slippery. "Did you become a priest before or after the illegal investment schemes you and Wissler dreamed up?"

Owling reached behind him, and the sound of bubbling started up. He pulled a hose to his lips and took a hit from the hose, then blew out a thick cloud of smoke. "The Navajo have a proverb which says, '*ma 'ii át'ah, ma 'ii dichin.*'"

"That's not entirely accurate, but I think you mean, 'coyote is always out there waiting, and coyote is always hungry.'"

Owling's eyebrow repeated the micro-raising movement. "You have dark hair like a Navajo. But Navajos don't have purplish-blue irises like yours."

"My paternal grandmother was Navajo." Drayco adjusted himself in the bag to keep from falling off. "But that saying doesn't explain your turn to crime."

"Doesn't it? Things don't fall in your lap. You gotta get out there and grab your opportunities."

"How did you and Wissler meet?"

Another puff from the hookah, another cloud of smoke swirled through the air. "Funny story, that. It was at the laundromat. We were both broke. And both breaking into the washers and dryers to steal coins."

"How long ago was this?"

He thought a moment. "About thirty-five years now." He chuckled. "Hard to believe. Where does time go?"

"Seems like you'd end up fighting over the spoils instead of teaming up."

"You'd think that, wouldn't you? But Stu, he was a charmer, that one. You wouldn't believe it to look at him. Rather plain. And those ugly glasses. But he had a way with words."

"And apparently, a way with women. Or a way with luring women into his schemes. Maybe willing, maybe not. Possibly a Russian woman or two, or women pretending to be Russian."

"Prostitution, you mean? I don't know nothing about that." Owling puffed away in silence for several moments, accompanied by the sound of the hookah bubbling and the recorded guitar music. "Sounds like something he'd do, though. Must have been after we parted ways."

"When was that?"

"Long time ago. Roughly twenty-five years, give or take. Oh, we bumped into each other now and then. But we'd formed different ideas of what being a hungry coyote means."

"How different?"

"I was okay with the money schemes. Stealing from the rich to give to the poor—as in me. But Stu was hooking up with some unsavory types. I didn't want to end up with concrete shoes at the bottom of Chesapeake Bay."

"Organized crime?"

"Loosely connected. More of a wannabe. I warned him not to get in too deep. Guess he must have listened since he never went to prison. Lost track of him the past few years."

"He's currently in little pieces at the bottom of the Atlantic."

At Owling's suddenly wide eyes that made him look like his Hooty-owl namesake, Drayco added, "Not the mob. Had a heart attack and was cremated. When the funeral director asked if he wanted anyone to participate in the ash-scattering ceremony, Wissler said the only person he could think of who'd love to see him end up as dust was Hooty."

Owling took a long puff of the hookah. When he exhaled, he made perfect smoke rings. And then he laughed so hard, his cheeks turned red. "He's probably correct, there."

"Your parting estrangement must have been worse than you let on."

"Oh, I hated him, all right. Loathed the man. A textbook narcissist, if there ever was one. He thought he was this slick, master criminal kingpin, but far from it. We were thrown together for a while, strange bedfellows and all. He knew I hated him. But he didn't care."

Drayco had to shift again to avoid rolling off the bean bag. *Note to self—never get a bean bag.* "Before you parted ways, you didn't hear him mention the names Lara Davidenko or Minna Hallow?"

"Sorry. But he and I had a lot of girls, if you get my drift. An eye for the ladies. Hard to recall all their names. I mean, we were stoned or drunk half the time. Speaking of women," Hooty turned around to put the hookah pipe back onto a holder. "If you'll excuse me, I have to get ready for my date."

"I'm sorry. . .your date?"

"As I said, I have an eye for the ladies."

"But you're a priest."

Owling cackled. "Episcopalian, not Catholic."

When Drayco left the building, he half-expected his Starfire to be gone. But instead, it was positively gleaming with a few beads of water on the windows. He stared at it, and Reginald sauntered over. "Looked like she could use a bath. Hope you don't mind."

"No, I'm grateful." He pulled out his wallet, but the other man put up a hand. "Keep your money. I done a few things in my life I'm not proud of. Father Owling says I can make up for it by helping others. Don't know if there's a heaven or not, but don't really like the idea of the other place."

Drayco nodded back toward the building. "About Father Owling. . .he seems unusual for a priest."

"Been in the gutter. Knows what it's like. Does charity work for types like me. And we don't forget."

"Good to know." Drayco opened the door to the Starfire, still marveling at how clean it was. "Thanks again for the car. She was getting a little dusty."

As he drove off, he mused on his day with the sinner-saints—a reformed drug addict monk by the unlikely name of Monk, who was an ex-Wall Street type, and an Episcopalian priest by the even unlikelier name of Hooty, an ex-con man type. Both were former grifters in their own way. Both were tied, directly or indirectly, to Lara, who'd allied herself with the kingpin wannabe, Stuart Wissler, and a commune filled with blackmail, jealousy, and squabbling rather than peace and brotherhood.

As the case wore on, the more he felt connected to Lara, née Minna. Yes, she was a schemer and a blackmailer, but she'd started out her young life motherless, like Alisa Saber. . .and like Drayco. She must have felt like an outsider, always trying to fit in, yet not fitting in anywhere. Small wonder she had big aspirations and grabbed the first ticket out of town in the form of Stu Wissler.

22

Friday, September 25

Drayco's dreams weren't filled with falling into pits of water this time, but of rampaging, tie-dyed motorcycles, so he wasn't at all sorry when he awoke at four. Sleep hadn't been his friend for several years now. If it wasn't nightmares, it was the hallucinations and sleep paralysis that happened right as he was waking.

No wonder he was addicted to Manhattan Espresso sodas. And salted coffee. As he made his way to his kitchen, he stood staring between the coffee pot and the refrigerator. Cold caffeine or hot? He opted for hot.

After he was fully caffeinated and dressed—and after a little Brahms at the piano to get him going—he spent time sorting through the mental notes from his strange, but not entirely unenlightening, interviews yesterday. He was so engrossed in that and more research into Alisa's case that the morning flew by. He barely had time for a quick bite of a three-day-old chocolate muffin before making his way to his Starfire, which was still gleaming. He should hire Reginald to wash his car on a regular basis.

He drove to his target, picked up his passenger, and they started the trek from the District to Silver Spring. His passenger, Alisa Saber, spent the entire drive over admiring Drayco's Starfire. "This is one cool car."

"Classic, yes. Doesn't have modern amenities. No DVD screen, no automatic parking, no dashboard that looks like a spaceship." He feigned a look of horror and said, "Whatever did people do before technology?"

She ran her hand along the dash and played with the toy piano mounted on top. "It's got AC. That's the only tech I care about right now."

They'd had to wait until early afternoon due to Alisa's morning classes at Georgetown. But he'd promised to help her track down her birth mother, and this was their best lead. They parked next to a small brick Cape Cod-style home, architecture popular in the area in the '50s.

Alisa wasn't quite as impressed with this fossil and said, "I guess my cousin isn't a gazillionaire."

Olga Whitman was allegedly about Drayco's age, but her bleached hair and tanned-leather skin made her seem older. A sun-worshipper, no doubt, but a healthy-looking one with intense hazel eyes that scrutinized the new arrivals. She welcomed them into her house, which appeared ordinary enough except for the strong sulfuric aroma of recently cooked cabbage.

She grabbed Alisa by the shoulders into a big hug and then held her at arm's length to study her face. "Oh, *kotehok*. Maybe a little of my cheeks. And my smile. And look," she pointed to the side of her nose. "A mole like mine." Olga had a hint of a Russian accent, possibly similar to what Lara had used.

She practically pushed them down into two large club chairs and plopped down inches in front of them onto a metal locker that served as a coffee table. "I was shocked when you emailed me, Alisa. Right out of the blue."

"I appreciate you didn't just delete it."

"I almost did. Thought it must be spam. But a blood relative? That I could not ignore."

Alisa said, "You're adopted? Like me?"

"Brought from Russia when I was a child by my adoptive parents. I tried to track down my biological family later. Didn't know their surname was Taras until I got it from documents in my mother's desk." She bit her lip. "I am not proud of snooping."

Alisa's eyes brightened. "What did you find out? About your Russian relatives?"

Olga's face grew wistful. "My biological parents gave me up for adoption. And do not want to have anything to do with me now."

Alisa cried out, "Oh, I'm so sorry. That's terrible."

"My birth was an unplanned mistake, an 'oops.' My mother was poor. My father, a man of some standing, was married to someone else. Age-old story."

"Still. That's downright criminal."

"I did find one relative in Russia to talk to me. She had a little information about a female cousin who disappeared. Might be your mother. But, I don't want to get your hopes up."

"What did this relative say? Why did her cousin disappear?"

"There were whispers. Something about the sex trade, but no one wants to talk about it. This cousin, Tasha, I do not know her last name, was a child of sixteen when she vanished. But again, it's been almost thirty years. So long ago."

Drayco's ears perked up at that. Tasha was also the name of Harry's one-night stand. It was all circumstantial, but if the puzzle pieces didn't fit exactly, they were awfully similar to the holes in the picture he had so far.

Olga stood up. "I have something for you." She disappeared into another room momentarily and then reappeared with a couple of photos that she handed over. "My Russian relative—the one who would speak to me—sent them. The first is a copy I made for you. A photo of my mother and a few aunts. The second is an original of the cousin, Tasha. Right before she vanished."

Alisa studied the new black-and-white copy of Olga's mother and aunts, and then the yellowed and fading photo of Tasha as Drayco looked at them from the side. Olga's resemblance to her mother was clear. The sixteen-year-old Tasha, on the other hand, looked like she could be Alisa's sister. The young woman had chestnut-colored hair that framed her light brown eyes and pouting lips. She'd made herself up to look much older with scarlet-red lipstick, heavy eyeliner, mascara, and glittery pink eyeshadow. Her hands were folded up behind her neck in a seductive pose.

Alisa said, "What else can you tell me? What city did she come from?"

"My Russian relative says our family is from St. Petersburg. Known for its art, beer, and canals. But I do not have much else. Sorry." Olga got up to snatch another piece of paper, which she handed to Alisa. "This is the name and address of my relative. And now maybe your relative, too."

Olga reached over to put her hand briefly on Alisa's arm. "I understand wanting to know your history. It feels like being a tree with no roots. One ready to topple over in a strong wind. But learn from what happened to me. Be careful."

"You mean sometimes it's better off not knowing?"

"It can be easier to take flight without the baggage of your past weighing you down." She smiled at the younger woman. "But I am glad I met you. Call me anytime you want to talk. And we shall go out for lunch soon." She winked. "I know a Russian restaurant a mile from here. Their borscht is killer."

When Drayco and Alisa returned to the car, he asked, "Would you mind if I take the photos to have additional copies made?"

Alisa cradled the pictures in her hand like golden eggshells, valuable yet fragile. "I don't want to part with them."

"I'll take good care of them. I promise."

She reluctantly handed them over, and he placed them in his briefcase in the back seat. She asked, "Where are we going next? You mentioned two stops."

"I called to check with that retired adoption worker you'd contacted. She's waiting for us."

ಌ ಌ ಌ

Fortunately for Drayco and Alisa, the town where the ex-adoption agency worker lived was on the way back, one reason Drayco arranged the side trip. Alisa had agreed to the visit, but she kept twisting around in her seat and fiddling with her seatbelt. When they arrived at the even more humble, one-story brick house, she sat glued in place staring at the house.

Drayco said, "You don't have to join me. I can talk to her on my own."

Alisa unbuckled her seatbelt and opened the car door. "Shall we go in?"

The woman who answered their knock wore a cannula and a small oxygen pack. A giggling and then a crash from behind her made her turn and say, "Don't you be knocking over those blocks again."

She waved Drayco and Alisa in and then hurried over to a baby in a high chair with a stack of wooden alphabet blocks at its base. Oddly, in their random order, they spelled out N-A-M-E-S. Nanci Abrahams explained, "My post-retirement job is babysitting my grandchild here. Meet Deedee. Deedee, meet Mr. Scott Drayco and Miss Alisa Saber."

Drayco and Alisa sat next to each other on a sofa while Abrahams parked herself close to Deedee to keep at arm's reach. Alisa asked, "How old is she?"

"A year and two months." She followed the baby's gaze, which was steadfastly focused on Drayco.

He grinned, "Do I have something on my face?"

Abrahams laughed. "I don't know. She never does that. Must be your eyes. They're unusual. Blueish-purplish."

"I'm grateful you were willing to see us, Mrs. Abrahams. Alisa's already told you something about our quest."

"I was shocked to get her call, I tell you what. Like a lightning bolt on a clear summer's day."

"You're used to dealing with confidential adoptions, which I understand. And we don't want to put you in a legal bind, but anything you can tell us would be a big help."

"I'm an ex-worker, and it was a long time ago." She studied Alisa's face. "I'd say twenty or so years ago."

Alisa replied, "Twenty-five."

"Most of the confidential part comes with adoptions where we have a paper trail. Records. Names, addresses, dates. This one was different."

Drayco said, "Alisa's adoptive mother told her Alisa was literally dropped off at your agency."

"That's right. Only case I ever had like that. That poor woman. That's why I remembered it after all these years. Quite frankly, it was the haunted look on that woman's face that made me decide to give up the job and go work somewhere else. I recently retired from a state social worker agency."

She studied Alisa's face. "That woman's expression—it's frozen in my memory even today. You look a lot like her."

Drayco asked, "This woman didn't mention any names?"

"Not directly, no. But I can remember the man who drove her. Very impatient. Kept yelling at her, using the name Tasha. And she called the man Stuart."

"You're sure about those names?"

"I have a good memory. People are always commenting on it."

"You told Alisa the woman spoke with a thick accent and was hard to understand."

Abrahams nodded. "That's true. Afraid I'm not up on accents, but it wasn't Spanish. She just said to take good care of the baby. And she had tear streaks on her face. I don't think this was something she really wanted to do."

"You also mentioned a scrap of paper with words written down?"

"That's a fact. Didn't understand them, either." She reached around to a table behind her chair, grabbed something white, and then got up to hand it to Drayco. "That's it, right there."

Drayco turned the paper over in his hands. It had writing on one side and only two words, *Pushkinsky* and *Gerasim.* "I'm surprised you kept this."

"Something told me to. I hadn't thought about it in years until I heard from you. I drug it out of an old shoebox of things I brought home from my desk at the agency."

"And she didn't give you anything else at all?"

"No, and even that one note she'd folded up to put in my hand, real secret-like. Don't think she wanted this Stuart fellow to see her do it."

Little Deedee started bawling loudly at that point, so Drayco signaled for Alisa that it was a good time to leave. As they returned to the car, he connected the timeline dots from Mrs. Abraham's info. If Tasha, Harry's one-night-stand and potentially Alisa's mother, also had ties to the shady lawyer Wissler, that explained how Harry came to be targeted by Wissler and Lara, Harry's fake "Russian" wife. Harry was a double patsy all along.

As they drove back toward Virginia, Alisa stayed silent for most of the trip, still fiddling with her seatbelt. Finally, she said, "This woman, Tasha. She was in trouble. She needed someone to help, so why didn't my father help her?"

"Harry Dickerman told Benny Baskin and me he tried to find her after she disappeared."

"Who was this Stuart person, then?"

"A bad actor. A slimy weasel type who liked to take advantage of vulnerable women. In some cases, they were victims, and in others, partners-in-crime."

"Was Tasha *victim* or *partner*, then?"

"I'm not sure. But I wouldn't judge Tasha too harshly until we know more. There's always at least some saint in the sinner and some sinner in the saint."

From Alisa's pale face and trembling hands, Drayco decided she needed a pick-me-up before he dropped her off at her apartment. He steered the Starfire toward a cafe not too far from her place. "Let me buy you some coffee. You look like you could use something."

She followed him meekly to an outdoor table where she managed to find her voice long enough to order an iced mocha. After some caffeine started coursing through her veins, a little color returned to her face. "Thanks for bringing me along today."

"I hope it gave you some of your answers."

"But there are so many more questions."

"That's the way investigations work. The truth comes out, eventually."

She sighed and drained the last of her mocha. "Thanks for the coffee, too. My place is only about eight blocks from here. Think I'd like to walk back. I need the fresh air."

"Are you sure?"

"Very." She seemed determined, so he acquiesced.

After he'd watched her as far as he could until she disappeared around a corner, he sat back in his chair, musing. If Stuart Wissler really was dead, it was too bad because he seemed to be at the center of both of Drayco's cases, Harry's and Alisa's. Then again, if he *were* alive, Drayco might want to off the guy, himself. Men who preyed on vulnerable women were the lowest of the low.

Deep in his thoughts, he was startled when a man's voice called out his name. He looked up to see the acting CEO of Harry's business, Sloan DelRossi. The man sauntered up to the table, and Drayco said, "Fancy meeting you here."

DelRossi pointed toward a building across the street. "One of those satellite offices I told you about. Had to pop by for a face-time meeting. A new campaign we're working on."

"Ah. Advertising is like the expansion of the universe. It never ends."

The other man tilted his head. "I take it you're not a fan?"

"Of universes, yes."

Del Rossi laughed, but then his face grew more serious. "I hope that reporter wasn't too hard on you."

"Reporter?"

"She came to interview Harry a couple of months ago. Was one of those business startup deals trying to be the next *Forbes* or *Inc.* Think it was something like TradeBiz. Or TradeCo. Can't keep them all straight."

"Are you sure it's the same woman?"

"I may be a married man, but I'm not a blind one. At the risk of sounding misogynistic, I never forget a pretty face."

He went inside the cafe to grab a drink for himself just as Drayco got a call on his cell. It was from Lara's former Maryland neighbor, Jilliana Vaughan, who'd forgotten that Lara gave her a box of things to keep for her. "I don't know how it slipped my mind." She added, "You want it?"

He did, and he told her he'd drop by that afternoon. It was a five-hour round trip in traffic, which is why he'd preferred to fly himself and Sarg over the last time.

ᘓ ᘓ ᘓ

As it turned out, with the traffic, car crashes, road construction, and road-raging traffic, it was more like six hours. Thus, it was nearing eight o'clock when he returned to his townhome, box in hand.

After he'd looked for the stray cat outside and made sure her food dish was full, he moved the box to his coffee table and opened it, wondering if it held treasure or trash. That he was surprised by the contents was the understatement of the year.

He got up long enough to grab a pair of nitrile gloves, then reached into the box and pulled out two passports, one for Minna Hallow and the other for Lara Ekaterina Davidenko. He also found some videotapes and a wad of cash in various bills, along with the addresses of five different bank accounts under the Hallow name.

So much for that vow of poverty. But it would also give the police more fodder that Harry killed her for her "estate," depending upon how much was in those accounts.

Drayco picked up the phone to call Benny Baskin. He'd have to turn the box over to the Fairfax detectives, but Benny needed to see it first. Somehow, Drayco didn't think he was going to be Benny's favorite anything after this little bombshell.

23

Saturday, September 26

Drayco must have looked as tired as he felt because when Sarg entered the door to Drayco's office, he immediately handed over a coffee. "Quad espresso. You still get these for early morning meetings, right?"

Drayco grabbed the cup, "I'm touched you remembered."

"Heh. You're—"

"Touched, all right. Same stale jokes. You haven't changed much, either, save for a few more gray hairs."

"Really?" Sarg went over to a cabinet and flipped open the door where a mirror hung on the inside. He ran a hand through his hair. "Elaine bought me some of that men-only hair goop. Supposed to cover up those strays."

"I thought gray hair made men look distinguished."

"Wait until you get your first one, junior." He sat down in the swivel chair and propped his feet up on Drayco's desk. "Found some tidbits on that murder victim of yours. Via Russian contacts."

"And?"

"Lara Ekaterina Davidenko, a.k.a Minna Hallow, *may* have used the name of a deceased woman in St. Petersburg who died around 1900. Probably got it off the internet. Who knows?"

Drayco opened his desk drawer to grab another pair of gloves before he reached into the box from the victim's neighbor. Last night, he'd called Benny, who was as upset about the box's contents as Drayco expected, but Drayco was going to hold off releasing it to the

Fairfax cops as long as possible. He picked up the two passports and opened them to show Sarg the photos.

Sarg grabbed reading glasses from his pocket and leaned over. "The Hallow one I get. Her legal name and all. Why have one in the fake Russian name? Unless she used it to convince Harry of her 'true' identity before the sham marriage."

"And to convince other men, too. Except for Stuart Wissler. He knew her as little Minna Hallow back when."

"Then maybe he was the one who had her use the fake name."

Drayco handed over the piece of paper the adoption worker had given him, and Sarg asked, "What's this?"

"The woman who dropped Alisa Saber off at the adoption agency handed this note over along with the baby. Oh, and she had a thick accent."

Sarg looked at the note. "*Pushkinsky* and *Gerasim*. Pushkinsky sounds familiar."

"A district in both Moscow and St. Petersburg. Also a couple of Russian bridges, a square in Moscow, and a few train stations."

"That narrows it down. And Gerasim?"

"I was hoping you could help with that via those Russian contacts of yours. Maybe check the records again for a Tasha Gerasim, or some variation, from a town that matches the word on that paper. We also have the name of Alisa's cousin, who was brought here under the adoptive name of Olga Taras Whitman."

Drayco hopped up to grab the photo Alisa's cousin provided, pinned to the research board on his wall—next to the original photo of Lara and Harry as well as the de-aged Lara photo. While he was at it, he grabbed those photos, too, and handed them to Sarg to study.

"The victim and Harry, right? And this one looks like a younger version of the woman in that photo. A little pixelated, though. I'm guessing software was involved. You make this?"

"Took my best shot at it, yes."

Sarg tapped the photos from Alisa's cousin. "Who's that?"

"A sixteen-year-old Tasha, supposedly."

"Ah. Could you make copies for me? Could come in handy."

Drayco headed to his desk drawer and pulled out some photos to give to Sarg.

"That was fast." Sarg peered at the drawer. "You got a copier in there?"

"Thought you might need them."

"Where'd you get the passports, by the way? Via a magic wand, your wizard-ness?"

Drayco blinked at him. "Have you been talking to Benny Baskin?"

"No, why?"

"That whole wizard thing."

At Sarg's confused expression, Drayco said, "Never mind. The victim gave her neighbor in Maryland a box of 'trinkets' that also included these," Drayco tapped the videos on his desk. "Plus a codicil to her will signed by two witnesses and notarized. And this," he reached down behind the desk into the box and pulled out the wad of bills he flashed at Sarg.

Sarg's eyes widened. "How much money is that?"

"Fifty thousand. She also had a ledger with names and addresses of five different bank accounts we didn't know about. All under the name of Minna Hallow."

"Harry named the beneficiary in those, too?"

"Yep."

"Well, well. Wonder how much moolah is in those accounts?"

"I have to take all of this over to Benny's office to show him, and he can notify the Fairfax PD. Guess I'll find out after they've done their thing. But one doesn't usually keep five bank accounts with only five dollars in each, do they? I imagine it's a whole lot more."

Sarg barked out a laugh. "So your victim could have been secretly rich. Wonder how?"

"Blackmail, possibly. Or Wissler gave her his money before he croaked. Or she stole it from him."

"Let me take a wild guess. Those tapes right there could have blackmail fodder on them." He looked at Drayco skeptically. "And you're just going to wait until the cops get back with you—oh, in a few decades—to find out what's on the tapes?"

Drayco pointed to a video player on a table against the wall behind Sarg, who dropped his feet and whirled around in his chair. "You made backups, didn't you?"

"Digital files, anyway. Haven't looked at them yet. Up to a viewing? Don't have any popcorn."

Drayco turned on the monitor, called up the computer files, and randomly chose one of them. With Sarg sitting glued to the screen and Drayco leaning against the wall, they watched for a few minutes until Sarg said, "Ugh. I've watched enough of those in my lifetime. You have too, remember?"

Drayco paused the video. "Looking at porn films was never a part of the job I treasured."

Sarg tugged on his ear. "The girl looks underage to me."

"Possibly. The man I recognize, though. He's younger here, around thirtyish. Older than the girl, for sure."

"Not Harry Dickerman, right?"

"No, his name is Seal Hettrick. He's one of the members of the Gaufrid Farm commune. I asked a monk named Monk, an ex-member of the farm, about Lara's blackmailing proclivities. She'd turned those 'skills' on him, and he thought it plausible she might be blackmailing other commune members."

"And this Seal guy was one of her victims." Sarg leaned back in his chair. "Hettrick, Hettrick, where have I heard that name before?" He grabbed his cellphone for a quick internet search. "I remember now. A prosecutor who specializes in sex crimes."

"Seal Hettrick?"

"No, her name is Glenda Hettrick. Hotshot attorney who skyrocketed through the ranks." Sarg flipped through the search results. "And from some bio info, looks like Seal Hettrick is the prosecutor's father."

"That must make for awkward family reunions."

"Think she knows about her father's, um, tastes?"

"Not if Lara was blackmailing him."

"Right."

Sarg cracked his knuckles. "This Gaufrid Farm seems like it's full of snakes. An ideal Hollywood soap opera setting."

Drayco grabbed a box from the top of the cabinets. "Funny you should mention snakes." He opened it for Sarg to take a look.

"Gah." Sarg slammed the box lid down again. "Where'd you find that lovely thing?"

"In my car. At Gaufrid Farm, right as I was leaving."

"Is that a copperhead?"

"That's what Reece Wable and I gathered from reptilian mugshots."

"Somehow, I find it odd that one just happened to wriggle its way into your car. Was it the Starfire or the other one?"

"Starfire. The windows were cracked, so there's that. I also took it to a mechanic for a quick check. He didn't find any easy access points."

"Might not mean anything. . ."

"Yeah. Might not."

Sarg's face looked a little glum. "Your cases are much more interesting than what I'm getting. Ah, how I long for those early days."

"We sure had some fun, didn't we?"

"You said it. I'll never forget that Fogerty case with the skunk. Took us a week to get the smell out of our hair. Nobody wanted to be within a mile of us."

Drayco grinned. "Guess people would call this," he indicated Sarg and himself, "a bromance nowadays."

Sarg snorted. "No offense, but if you were the last person on Earth, I wouldn't be interested in a 'mance' with you."

That brought on a full-throated Drayco laugh, so much so he had to grab some of the coffee when it set off a coughing fit.

"Where you off to next, junior? Snake-hunting? Porn-sniffing? Russian-women dancing? Which you should know something about, I might add."

Sarg hadn't known Drayco's Russian fiancé personally since she broke things off before Drayco joined the Bureau. But Sarg had sometimes been on the receiving end of his partner's half-drunken ramblings early on in their early FBI days.

Ignoring Sarg's comment, he replied, "I'll have you know I'm off in search of an invisible man."

24

Was it only a couple of weeks since Drayco's last case, up in Pennsylvania? He was having a touch of déjà vu. After leaving Sarg and the District behind, it had taken a few hours due to the usual D.C.-area gridlock. But at least there weren't any late-summer storms to drive through on his way to Travilah. It was a relatively small unincorporated area in Maryland, not far across the border from Pennsylvania. What it lacked in size, it more than compensated for in income, making the top twenty-five list of wealthiest towns in the country.

Armed with some records research and the info from his visit with Darren Monk at the monastery, Drayco had tracked Ivon Leddon's origins to this very house. He studied it from the bottom of the diamond-shaped driveway. The place had hints of Frank Lloyd Wright, although Drayco hadn't found any actual ties to the architect. More homage than anything.

The building itself was in an L shape, and from the aerial views Drayco found online, it backed up to a small lake behind. It really was striking. Angular glass-lined corners with long yellow bricks flanked the exterior, and a white-walled balcony bisected the two floors all the way around the house.

After a maid greeted Drayco, she led him down a narrow hallway, and the house's charming eccentricities carried over to the room where she deposited him. He noticed an indoor greenhouse area—complete with a rock sculpture, waterfall, and pond. Floor-to-ceiling glass windows looked out toward a pool and the lake beyond.

The man who scampered in and shook Drayco's hand had the same smile and smattering of freckles as a photo Drayco had found of

Ivon Leddon from the man's trader days. But the two men were obviously not one and the same. Where Ivon was tall and had sandy hair and brown eyes, Porgy was several inches shorter, with dark hair, gray eyes, a Dalí mustache, and hint of a goatee.

Drayco said, "You're Porgy Leddon?"

"That's me. And you must be Scott Drayco, the investigator. I hope you've had word of Ivon?" The man bounced on his shoes.

"I wish I could bring some good news. But unfortunately, I'm like you, looking for answers."

Porgy sagged a little at that and gestured toward a low-slung orange leather sofa that didn't look terribly inviting. Hopefully better than a bean bag. "Please join me. I'll tell you what I can. Love to help. Yep, help's my middle name."

Drayco was surprised at how comfortable the sofa was. "This is an interesting house, architecturally."

"Isn't it? I was always jealous of Ivon when I used to visit. After his parents passed, Ivon sold it. When it came back on the market a few years ago, I couldn't resist. Just couldn't resist." Porgy sat down and then hopped back up again, opting to stand behind his chair, still bouncing on his shoes.

Drayco was bemused by the man's hyper behavior. Addiction, perhaps? To caffeine or cocaine?

When Porgy saw Drayco eyeing him, he chuckled. "My fiancé says I need to cut down on my habit. Love my quad espressos."

"I'm a bid of a caffeine-head, myself. Had you heard from Ivon much since he went into the commune at Gaufrid Farm?"

"Not all that often, with his mother and father gone, and all. We get along well but aren't ultra close."

"So he didn't try to contact you or other family right before his disappearance? To say he was going to be traveling? Or give any hint he might want to vanish on purpose?"

"Not a word, none whatsoever. I hadn't chatted with him for over a year. Don't recall anything weird from that call. Maybe he was more pensive."

"He wasn't afraid or upset?"

"Seemed his usual self. Yep, seemed quite normal. I've thought about that call since he vanished. Played it over and over again in my mind. But it was just ordinary."

Drayco looked around the room. "I guess being raised in a family with money is what spurred him on to get into trading and investing?"

"Ivon's always been a high-powered Type A personality. Trying to be the top of whatever he's into at the time. Climbing the Wall Street corporate ladder, raking in money, and making deals appealed to him. That's why I was so shocked when he decided to give it all up and go to this weird commune."

"Did he have any secrets that might have set him up for blackmail? There's a possibility other members of the commune were being blackmailed."

Porgy scratched his chin. "Ivon? Seems unlikely. He was a bit of a Boy Scout underneath it all. In fact, he made it to Eagle in his teens."

Drayco eyed him skeptically. "Are there badges for wheeling-and-dealing?"

"You got me." Porgy paused to look out toward a bronze sculpture with a green patina not far from the pool. He pointed at it. "Ivon made that. He's a whiz with tools. Can build anything, fix anything. Thought he'd end up as a contractor. Or artist or something."

"Did Ivon have any friends overseas he might try to visit if in trouble?"

"Overseas? I don't think so. After he went into that commune, his friends dropped off. And his New York connections. . .Wall Street isn't exactly the place to make friends. Just deals." Porgy narrowed his eyes. "Look, you can't believe Ivon left the country because he was on the lam. That's utterly ridiculous. A hundred percent ridiculous."

"Since there was blackmail going on at the commune, he might have been targeted and looked for an escape."

"That supposes they had something to blackmail him over. Maybe he forgot to dot some I's or cross some T's on a contract or something. He didn't have any vices."

"No drug problems?"

"Didn't even take aspirin when I was around him. Nothing illegal."

"What about relationships? Girlfriends, boyfriends, lovers?"

"He was straight. Never married. Dated a lot, as I recall. Had a boyish charm to him. Guess that appealed and all."

"Did he have a preference for younger. . .dates?"

Porgy glared at Drayco. "If you're asking what I think you're asking, nope. You are way off."

"Of course." Drayco pointed to a painting hanging over the fireplace at the opposite end of the room. "Is that another of Ivon's creations?"

Porgy turned to see what he was referring to. "That's a Serge Poliakoff. Can't afford a Picasso. Yet."

"You own a chain of dry cleaners in the state?"

Porgy beamed. "Added two new stores this past week. Thank god for wool and men who won't do laundry." He checked the time as he started bouncing on his feet again. "I've got a video meeting here in about fifteen minutes. Feel free to walk about the grounds or take a look at the sculpture."

Drayco got up to leave, but Porgy made him stay long enough for a more thorough tour of the house, including the indoor waterfall and pond that actually had koi. Drayco also accepted Porgy's invitation to stroll around the grounds and started with Ivon's sculpture, as the man suggested.

It resembled a womanly torso in a Henry Moore sort of style. It was pretty good. He studied it, hoping it might send him some clues, but nada. A creative, if driven, soul who ended up on Wall Street and then in a commune—Drayco couldn't reconcile the various warring factions of Ivon's tastes with his background and life choices.

Porgy was flat-out sure his cousin wouldn't have sex with underage "dates," nor did he allegedly do drugs. If so, he didn't suffer from the same vices as Seal Hettrick and Daven Monk. Perhaps blackmail wasn't the cause for his vanishing act at all, unless Drayco's earlier musings were true, that Ivon was a partner with Lara in the blackmail, and she killed him. Or he killed her over it and had to disappear.

If not blackmail, then what? One thing was clear—people associated with Gaufrid Farm had a strange way of disappearing. Or dying from drug overdoses. . .or murder.

25

He would have loved nothing better than to pick up some takeout from SiAm Thai Emporium and head on home, but Drayco had forgotten his briefcase at his office, so he made a quick detour. He found a parking space alongside the venerable brownstone and hiked up the three flights of stairs.

Everything seemed all nice and normal—except for the two men who were slinking out of his office. He was pretty sure they weren't maintenance workers, even though they wore gloves.

He called out, "Hey!" as he ran toward them.

The first one, built like a gorilla with a buzz cut, looked like he was going to make a run for it, but to do so, he'd have to get past Drayco. The second one, with a stance more like a welterweight boxer, narrowed his eyes with a snarl and threw himself at Drayco in a full run.

Having anticipated such a move, Drayco sidestepped the much-heavier man and gave him a karate kick as the man hurtled on past. Goon Number Two went down in a heap while Goon Number One lunged at Drayco and grabbed him, vise-like, from the front. Drayco kneed him in the family jewels, and when the groaning man let him go, Drayco gave him an uppercut with his fist. Unfortunately, the effort knocked Drayco off balance.

Goon Number One called out, "Anikey, *иди к машине*. Vienna," and his fellow goon companion managed to stagger up and hurriedly join him as the duo fled down the stairs.

Drayco had to make a split-second decision whether to check on his office for fires, explosives, or other hazards the goons left behind or to pursue the goon squad. Since he didn't smell any smoke, he opted for the goon squad. By the time he scrambled down the stairwell, ignoring the pain from a pulled quad muscle, they were out of sight.

But where? They hadn't exactly shown they were that nimble. Then he heard the sound of a revved-up car engine and spied a dark pickup speeding away. He got the color but not a model and only two of the numbers on the plate.

Seeing there was no chance he'd be able to jump into his Starfire and catch up with them, he limped back to his office, expecting the worst. It was bad, but not quite as bad as he'd feared. Papers lay strewn around the room, with desk and cabinet drawers open, and it was apparent they'd tried to hack into his password-protected computer. No explosive devices, thank god.

With a sigh, he went about taking photos and lifting prints that he'd get one of his MPD friends to run tomorrow. Probably a fool's errand, seeing as how the goons wore gloves, but he'd be remiss if he didn't try.

Were these the same thugs responsible for ransacking Lara's apartment? If so, had they planted bugs here like they did there? He grabbed his RF detectors and did a complete circuit of the room. No radio bugs. But did they find what they were looking for, was the question. Nothing seemed to be missing. . .until he looked up at the case research board on his wall.

The yellowed photo that Alisa's cousin had given her of Tasha as a teenager—the one Alisa had entrusted to him—was missing. And yet, Lara's photos were still there, both the snapshot with Harry and the de-aged one he'd made. Why? What could they want with the photo of a teenager taken some twenty-five years ago? Maybe they'd confused the two photos, although Lara was seven years older in her picture than the younger Tasha.

One thing was certain. . .the two men spoke Russian. From the exchange between them, Drayco thought he might have a partial name and a potential destination where they were headed.

He'd also got a whiff of leftover tobacco smoke that had a chemical tinge to it, possibly from Russian Belomorkanal cigarettes? Few places sold those, something he could try to trace. And there was an unusual tattoo on one of the goons, of a cat wearing a hat and definitely not the Seuss character. Hardly a lot to go on.

Still mulling over that and the encounter, he was on high alert as he drove to his townhome. Even more so when he opened the front door and heard noises coming from the kitchen. He paused and listened for a moment. The noises hadn't seemed to stop when he entered, so he grabbed a heavy flashlight from the table beside the door and tip-toed around the corner, flashlight lifted over his head.

He dropped his arm as soon as he saw his "intruder" and placed the flashlight on the counter. "I see you're helping yourself to my coffee again, Dad."

"You always have great coffee. Where do you buy it again? Compass? Regarding food, on the other hand," Drayco's father opened the fridge and peered inside, "not so much."

Brock doctored his coffee with some cream and sugar and nodded at the flashlight. "Power's not off. Wanna tell me why you seemed ready to attack your poor gray-haired padre with a blunt weapon?" He strolled over and picked up the light. "Damn thing's heavy. Good skull-cracker."

Drayco grabbed a mug and poured himself some of the coffee. "Had a run-in with some goons at my office."

Brock blew on the hot liquid and gingerly dipped in a pinkie. "Goons? What kind this time?"

"I overheard them say a few words in Russian. And with all the Russian links I'm turning up in my current case, maybe I should watch out for a ricin umbrella jab."

Brock frowned. "That's not very funny."

"Isn't it? I can't tell anymore. My funny bone is broken."

"When you told me about your case over the phone, I was a little concerned. Sounded like more of an international crime syndicate thing. Best left to our former colleagues at the Bureau."

"Sarg has offered his assistance. That should count, right?"

"Benny Baskin has enough evidence to form a plausible case to show this Dickerman guy is likely innocent, yes? So why not drop this one? Must be awkward with Darcie involved."

That was the third time. . .or was it the fourth?. . .someone had brought up the subject, and Drayco was getting sick of it. "As if I couldn't be objective, is that what you're saying? That I'm not professional enough to treat this like any other client?"

"Just worried, that's all. You've been put through the wringer with your last several cases. That suicide case, and your mother. . ."

Drayco talked himself down off his testy high-horse when he realized his father was being over-protective, something Drayco hadn't seen that much of in the past. It was kind of nice. Even though it was also kind of annoying.

Brock pointed to an empty bag of cat crunchies in the trash. Like his son, he was also a master of changing the subject when the emotional heat got turned up between them. "You'll need to buy some more. I poured the last of those into the food dish out back."

His father had remembered to feed the stray cat? That was unexpected. "Did you see any signs of her?"

"She's pretty shy when I'm the one to take on crunchy and water duties. Still skittish."

"I was hoping she'd come in long enough for me to call a rescue group. They could find a foster home for her."

Brock topped off his coffee. "You should keep her."

"Me? My schedule is irregular. Wouldn't be fair to any animal."

"You do fine with my Hoover pup."

"Yes, but you're the one taking care of him. Which reminds me. You never said why you name everything after that disgrace to the Bureau. J. Edgar Hoover abused his authority. Crossed all kinds of ethical and legal lines."

"How do you know I didn't name the dog after Herbert Hoover? Or the Hoover vacuum company? Or Willis Hoover, the country and western singer?" Brock added, "Besides, you've got Hoover in *your* name."

"Not many people know about that extra middle name, and I like to keep it that way."

"Suit yourself."

Brock studied Drayco's face, and Drayco had felt the older man's gaze on him the entire time he padded around the kitchen. So he finally said, "What? Did I grow a third arm? Which would be pretty awesome for piano playing, I have to admit."

"Just checking. Don't see any bruises. You're limping, but only a little. Guess you had the upper hand with your 'goons.'"

"They weren't Olympic athletes. I'm fine."

"Make sure you stay that way. Don't forget I'm hosting Thanksgiving at my house this year. It's only two months away. I'm inviting Benny Baskin and his wife and the Sargosians, if they can come. Plus other assorted folks. Of course, you'll play the piano for us."

Drayco groaned inwardly. He did not enjoy parties. And he didn't like being put on the spot for a "concert" for a gathering without enough prep work. He'd have to soak his right arm for an hour beforehand. If he were lucky, maybe an asteroid could strike the Earth right about that time.

Brock rummaged around in the cabinet and pulled out a bag of lemon cookies with a triumphant cry. "All is not lost." In-between munching, he said, "This case of yours. Sounds pretty complicated."

"No more so than most. A few of the major players are dead or missing, which makes it more challenging."

"You'll figure it all out. But a commune for ex-Wall Street wackos? Seems fishy."

"Gaufrid Farm. Name's derived from old German and means land of peace."

Brock leaned against the counter. "I'd say it's the opposite."

"You won't get an argument from me on that." Drayco grabbed a lemon cookie and dunked it into his coffee. Not bad. And they were only slightly stale.

26

Sunday, September 27

Drayco tried not to yawn as he headed down the Pacific Building hallway toward Benny Baskin's office. Brock had stayed until midnight the previous evening, thanks to the new basketball hoop he'd convinced Drayco to install in his backyard in a weak moment. Maybe that's why the feral cat didn't come around when Brock was there.

Just as Drayco reached for the doorknob, he got a call on his cell. He looked at the caller ID and backed up toward a bench down the hall. "Sarg, what have you got for me?"

"That's rather presumptuous of you. Perhaps I was merely calling to ask about your health. Or the weather. Heard there's going to be a bad storm this weekend. Hail, even."

"Thanks for the weather report. So again I say. . .what have you got?"

"Think we've tracked down the Tasha person. Her name is Tasha Oleneva. Came to the U.S. on an H-2B Visa at age sixteen from St. Petersburg. The visa lapsed one year later, and there's no record of renewal or her leaving. Nor are there any records of her after that date back in Russia, although those are more spotty. Like she vanished into thin air."

"More like she fell into the clutches of Stuart Wissler."

"About that guy. He wasn't on the Bureau's lists in the sex trafficking division. Or the State Department's or Homeland Security. He also didn't seem to be officially connected to organized crime circles, for that matter. Doesn't mean those ties don't exist. Just that he was too small a blip to register on anybody's radar."

"Thanks, Sarg."

"*De nada*. But I'm not forgetting you owe me a seafood dinner after the one in Maryland was a bust."

When Drayco finally made his way into Benny's office, he was happy to see Nelia Tyler, too. Neither Benny nor Nelia was usually there on a Sunday, but Benny had grabbed the chance to get out of his wife's landscaping schemes. Nelia, on the other hand, had worked for Sheriff Sailor Friday and Saturday instead of her normal weekend slate.

Drayco filled them in on Sarg's call, and Benny's reply echoed Drayco's thoughts. "If that Wissler guy was involved with trafficking, it's too bad he's dead. Because I'd like to kill him, myself. Traffickers are the lowest pond scum on the planet."

"I'm leaning more toward a con man or trickster who preyed upon vulnerable women as his partners-in-crime."

Drayco slid into the hard wooden chair to the right of Benny's desk, clearing the way for Nelia to sit in Drayco's favorite Sangria-colored leather chair. Benny started pacing across the room, barely missing Drayco's chair by inches. "Okay, kids, what have we got here. Blackmail?"

Drayco nodded. "Daven Monk and his cocaine. Director Gordon Aronson's secret wealth. Seal Hettrick's pornography with potentially underage girls. Catherine Cole's jealousy regarding Aronson's love for our victim. Max McCaffin's money laundering, and maybe drugs, too, since one of the former commune members OD'd."

Benny stopped his pacing long enough to ask, "What about the commune members' former businesses. Blackmail fodder there?"

Nelia looked at the notes Drayco had given Benny. "Most of the current members once worked for companies with ethical problems, fines, or sanctions. Max McCaffin's Davos Electroplating Services, a subsidiary of Theunissen Trading, was fined for dumping and fraud. Catherine Cole's Abler, Macedo & Ochs Financials got dinged for defrauding investors. Seal Hettrick's Kiedaisch Investments had a conviction for insider trading. Gordon Aronson's Bachert Brothers was cited for securities violations. Though some of those date back a quarter of a century. Especially Max and Gordon."

Benny frowned. "What about the missing man, Ivon Leddon?"

Drayco said, "Far as I can tell, his former company, Zigow Equities, was pretty clean. And the man has no rap sheet of any kind. Plus, my chat with his cousin, Porgy, indicated he was a regular Boy Scout type. Literally."

"Okay, so what about Harry's daughter and her motives?"

"That's murkier." Drayco adjusted himself in the hard wooden chair. He was getting so tired of uncomfortable chairs, he should go into furniture design. "She may have posed as a reporter to investigate Harry before Lara's murder." Drayco added helpfully, "You'll be happy to know the commune member-turned U.S. Representative, Boyce Hershorn, has a rock-solid for the time of death."

"Gratias Deo I don't have to deal with a congressman on top of it all."

"I'd think you'd relish the challenge." Drayco had to stifle a chuckle recalling that Sheriff Sailor had the same reservations as Benny.

"Oh, it's not so much I hate taking them on because I'd love to. But I loathe all the added publicity it brings to a case."

Nelia flipped through the notes again with a puzzled expression. "But why blackmail in the first place? If the commune members took a vow of poverty—despite Gordon Aronson's alleged stash—then how would Lara be able to squeeze blood money out those turnips? And for what purpose? Or is this all a money-laundering thing?"

Drayco answered, "Lara's neighbor said she wanted to go to Russia or some other country and start a new life there. That might be the purpose. As to the money. . .you're right about Aronson. I found out yesterday the guy owns property in a couple of different states. And Wall Street *is* the center of the money-laundering universe."

Benny was still pacing. "Did I tell you, Drayco? I got word the Fairfax PD followed up on the victim's bank accounts from that box of yours. She'd deposited the funds over the past several years in cash. She once joked with a teller that it was a regular payment from a friendly relative. Didn't say which one."

"According to the town librarian where Lara grew up, she didn't have any living relatives left here."

"All of that may not matter since there's one other teensy problem with those bank accounts."

"You mean the codicil naming Harry beneficiary of them, too?"

Nelia shot him a sympathetic look. "The Fairfax PD were delighted by that news."

Drayco shook his head. "As I've said before, Harry's already rich. I'm not seeing it as a motive for murder."

"You know cops. They love the easy way out. File away one case as quick as you can because there are plenty of other folders waiting. But you did shame them into testing that tree in Harry's backyard."

"They found something?"

"It rained since, but they did find a small blood sample on some interior bark. It's not the victim's. Nor does it match anything in databases. They said it could be anyone, even a tree trimmer. They did grudgingly admit it's puzzling it would be there."

Drayco replied, "A squirrel assassin taking out the chippies."

"Sarcasm doesn't suit you, boy-o. Now, what's all this about an incident at your office?"

Nelia turned to him with raised eyebrows. "An incident? When?"

"Yesterday evening. A couple of goons speaking Russian went through my office. I caught a partial name and have alerted the area PDs. Plus, Sarg knows about them. I also took some prints just in case."

Nelia asked, "Russians, as in transnational-organized-crime Russians?"

"If so, the standards are slipping. These acted more like cartoon villains."

"Still, that snake and now thugs. You'll be extra careful, right?" Her face scrunched up in a wreath of worry-wrinkles.

Benny gave Nelia and then Drayco a sharp look in turn. "Drayco's always careful, aren't you? Careful's his middle name."

It was actually Scott Ian *Hoover* Drayco, but he wasn't about to let Benny know that. Drayco said, "Cheer up. I think it's doubtful Harry hired goons from his jail cell to prevent me from proving his innocence."

Just then, Benny got a call from his wife and made yakking motions with his hand. Nelia motioned for Drayco to join her in the hallway outside. She studied his face. "Are you sure you're all right? I noticed you limping."

"It's nothing. A little pulled muscle."

"If you're sure." Then she formed a slight smile. "You got the best of two thugs, did you?"

"In a manner of speaking."

"Still." She folded her arms across her chest. "Sheriff Sailor would be proud. He always did say you'd make a great deputy."

"Sailor said that? Oh, now I'm going to have to bake him some cookies."

She shorted. "Pie, you mean."

"How could I forget?"

He caught a whiff of perfume, a bit citrusy. She didn't usually wear perfume. When he pointed it out, she uttered an embarrassed laugh. "Since I'm running around so much, I was afraid I'd start to 'offend.'"

He blurted out before his mind could warn him, "And here I was thinking you'd worn it just for me."

Her eyes widened, and she coughed but recovered enough to say, "If a girl wanted to attract a man in your business, she'd have to use 'eau de ninhydrin fingerprint solution.' Or hexanal."

"Add in a crime-scene-tape scarf and a pair of magnifying-glass earrings, and I'm hooked."

She bit her lip, although whether it was to hide a smile or embarrassment, he couldn't tell. But the charged air between them was still a little *too* charged as it often was, so he asked, "Who are you putting your money on among our various suspects?"

"I'm liking Seal Hettrick right now, what with those tapes."

"A porno fan, I see." And he immediately regretted those words, too. What was wrong with his runny mouth? He wasn't ordinarily this. . .teenager-ish.

But he was relieved to see her eyes finally light up with laughter as she replied, "I hear you can find a lot of that on something called the internet."

"Yes, I believe I'd heard of that, too." He tilted his head. "Wonder if they have any piano porn?"

She broke out into a fit of giggles, and he determined to add "piano porn" to his online research later. Anything that could get that kind of reaction out of the usually serious deputy was worth a look.

27

Drayco was in a better mood than he should have been after leaving Benny's office and Nelia. But that didn't last long as he headed toward a meeting with Alisa Saber. Truth be told, he was angry about being lied to.

He'd asked her to meet him at the same cafe where they'd had coffee two days ago. Not only because it was within walking distance of her apartment, but it was the same location where he'd bumped into Sloan DelRossi, acting CEO of Harry's company.

As he approached, he saw the young woman sitting at an outdoor table and holding a menu as the sun glinted off her flaxen hair, creating a halo effect. Was she really an innocent "angel" or something else entirely?

She'd just ordered a turkey-on-rye and jasmine iced tea when he sat down. She said, "I can call the waitress back."

"No need. I'll just have some coffee. I had a late breakfast."

As he continued to stare at her, she laughed nervously. "You look so serious. You're not a vegan, are you?"

"Are you familiar with Harry Dickerman's company, Mediasio?"

"I think I've heard of it. Why?"

"The acting head of that company saw you and me together at this very cafe. He told me afterward you were a reporter who'd interviewed Harry a couple of months ago. Allegedly from a business startup deal called something like TradeBiz or TradeCo."

"He said that?"

"Funny, though, I did some research into all the various business publications, new or established. None has an Alisa Saber listed as an employee, reporter or otherwise. Nor did I find anything close to TradeBiz or TradeCo."

She twisted the cotton napkin in her lap. "Okay, so I might have done something like that and didn't tell you. And maybe it wasn't entirely ethical. But it's not what you think."

Drayco leaned back with a frown. "Want to tell me more?"

"I was trying to figure out what kind of man Harry was. Before I went any further with this paternity thing. I didn't want to set myself up just to be crushed if I found out he's a louse. Although if he's a murderer, I may be too late for that."

She paused while the waitress brought her drink, and Drayco ordered his coffee, black. Perhaps it was her acting training, but as she hunched over her drink with a pinched forehead and pallor to her cheeks, cracks in her self-confident façade broke through for the first time.

He asked, "Where were you on Friday two weeks ago, around the time Lara Davidenko was killed?"

Her mouth opened and then closed with an audible snap. Up went the shield again. "Then you actually think I could be guilty of killing that woman?"

"So, where were you?"

"What was the date again? And time?"

"Friday, September the fifteenth. Around eight o'clock."

She sat back and thought for a moment, then grabbed her cellphone from her purse and scrolled through a calendar. "I have a class. One of my few evening courses. So it would have been impossible for me to leave, kill that woman, and get back to school. You can check with my chemistry professor about whether I was there or not. It's Dr. Peter Heyer."

He nodded. "I'll look into it."

"Please don't tell the cops."

"I may not have to, due to those client privileges you asked me about. But if I determine it's relevant to the investigation, I'll have no choice."

Her sandwich and Drayco's coffee arrived, but she poked at her food and didn't seem particularly interested. He reached for the salt shaker to pour some grains into his coffee, which got her attention.

"What in the world are you doing?"

"You should try it. Better than sugar."

"You're insane."

He smiled. "I've been told that before."

She picked up the sandwich and tentatively took a bite. It must be unusually good or she was unusually hungry, because she finished the first half off in a couple of quick bites.

"I haven't forgotten about the other reason you hired me, Alisa. I've got a few possible leads on Tasha, the woman your cousin mentioned."

"Did you find her?" Alisa's face brightened with a glimmer of hope he hated to dash.

"Not yet. Her full name may be Tasha Oleneva from St. Petersburg, Russia, who arrived here on a visa at age sixteen. But the visa expired, and she disappeared."

Alisa sounded out the name slowly. "Tasha Oleneva. It's kinda musical, isn't it?"

"It is, although I think her interests lie more with modeling than music."

"She's very pretty, isn't she? I mean, that photo."

"She is that."

"But by modeling, I guess you don't mean a Paris catwalk."

"That might have been her goal, hard to say. But I believe reality proved far different. And not nearly as nice."

Alisa finished most of the rest of her sandwich and pushed the plate away. "I can't imagine being in her shoes. Leaving your country so young. Hoping for something big. Then ending up in some sleazy relationship leading nowhere."

Drayco opened up the briefcase he'd brought and handed her a photo. "That's a new copy of the one you loaned me of Tasha as a teenager. The original was stolen."

"Stolen? What do you mean stolen? You told me it would be safe with you." Her face flushed a bright red, and her eyes had a fire to match.

"Two men broke into my office and took it, and I don't know why yet. But I will. Meanwhile, I'd scanned your photo into the computer. I cleaned it up and printed it out on photo paper."

She studied it. "Looks the same. If not a little better. I still want the original because my cousin gave it to me. And it came from my other relatives."

"I'll get it back. Don't worry."

"You wanted me to trust you. And I did. But, if you were wrong about that, why should I?" Her eyes were positively blazing now.

"I didn't anticipate this turn of events. But it may be a blessing."

"A blessing? How is this a blessing?"

"The two men who broke into my office spoke Russian. And I may be able to track your mother through them."

"Oh." She looked down at the photo. "If it will get me closer to being reunited with my mother, then okay."

Like the last time, Drayco watched as Alisa left the cafe and walked back toward her apartment, keeping an eye on her as long as possible. Would he be able to keep his promise to find the stolen photo?

She had every right to be angry. Hell, *he* was angry. He only had a few photos of his mother, and if someone stole them, he'd be pissed, too. He stifled the urge to pull out Maura's picture from his wallet again—to make sure it was still there.

When his phone chirped out a text alert, he cursed it. What would it be like to have a minute or two without technology interrupting every aspect of life? When he read the text, he rubbed his forehead. One of the feelers he'd put out had resulted in an invitation, thanks to Brock, who had a few more powerful political connections than Drayco.

It appeared Drayco was heading to Congress. He might live not far from Capitol Hill, but he was much happier jogging by the place than he was diving into the belly of the beast, surrounded by all that bile and hot air. It couldn't be helped, because he really needed that appointment. But it didn't mean he wasn't half-dreading it all the same.

Brock was a lot more comfortable in that realm, but Drayco's father was more at ease with schmoozing in general. For Drayco, life was too complex, too *urgent*, to condense it to inanities, fake smiles, air kisses, and idle promises. He'd be more at ease in a real lion's den where you knew exactly who your opponents were.

28

Monday, September 28

It wasn't the first time Drayco had stepped inside a politician's office on a case, although this particular representative, Boyce Hershorn, was harder to contact than most—considering he was a freshman rep. It took the extra nudge from Brock to help since Drayco's father knew Hershorn from one of his own cases, and they'd gone golfing together once.

Hershorn's office could have posed for a stock photo, with the obligatory state and American flags, and gold-framed copies of the Declaration of Independence, Constitution, and George Mason's Virginia Declaration of Rights on the wall. The only personal touches included a family photo on the desk beside a "Virginia is for Lovers" mug and a dish filled with wrapped candies stamped with "V."

He rose to shake Drayco's hand when Drayco was ushered in. "Brock's boy. I can see the resemblance. A couple inches taller. Please, sit down."

"Thank you, sir. I'll try not to take too much of your valuable time." Hershorn chose to stay standing, so Drayco did the same.

"This is about Lara Davidenko, you said?"

"Yes. Have the Fairfax police spoken with you?"

"Actually, they have not. Although I was half-expecting something like this when I read about it in the papers. Guess they knew I was on the floor of the House for a late vote at the time. Ergo, not a viable suspect."

Drayco had checked on that himself, accessing an archive of the televised proceedings. "I've spoken with the remaining members of

Gaufrid Farm and Daven Monk. Most of their former employers had a record of shady practices. Makes membership at the farm seem more like a salve for a guilty conscience than looking for utopia."

Hershorn's lips curled up into a brief smile. "You're wondering if I had a similar stain in my background?"

"You departed a job at Merced & McClung, a big corporate law firm for major trading companies. You were even a partner at the time. I'm curious why you left."

After a moment's hesitation, the congressman replied, "I'm assuming this is a confidential discussion?"

When Drayco nodded, Hershorn continued, "I suppose I fit that commune profile of yours. Joining had to do with guilt, probably like the others. The greed, the drugs, the hypocrisy on Wall Street. . .it gets to you."

"Why leave the commune, then?"

"When I'd had enough of lashing myself with the whip, I decided I could do more good through public service. A different form of penance."

"How did you hear about Gaufrid in the first place?"

Hershorn shrugged. "From friends. I was on the verge of a nervous breakdown at the time. My failing marriage, the pressures of the firm, the long hours, my infant son passing from SIDS. Guess I wanted to run away, and the place sounded ideal."

"But it wasn't?"

"After four years, I realized I couldn't escape from my problems. Decided to start up my own law firm. Even then, I realized *that* wasn't enough to help people. I needed to be able to represent them in congress."

Drayco glanced at the family photos on the representative's desk. There seemed to be one missing, a familial link he'd just discovered. "Are you sure it wasn't Catherine Cole who gave you the recommendation? Seeing as how she's your sister?"

Hershorn rubbed his forehead. "She was a big part of my going there, yes."

"Have you been in touch with her recently?"

"When I heard about Lara's murder, naturally I was worried about Catherine. I got word back to her to make sure she was okay."

"Why do you hide your relationship? Director Aronson doesn't even seem to know. He told me her family lives up in Maine, and she never talks about them or visits."

"Gordon prefers not to have family members join together. Plus, Catherine is my half-sister."

"According to my sources, from an affair by your father with his former maid."

Hershorn grimaced and side-stepped the comment. "My father helped her out by getting her a job at a trading firm where she acquitted herself well. Until she left abruptly to join the commune."

Drayco walked over to get a closer look at his copy of the Constitution. "Did she leave because execs at Catherine's former company were convicted of defrauding investors?"

"You'd have to ask her about her true motives behind joining. But what you say about her company is correct."

"You must have left on good terms with Gordon, seeing as how you authored a bill that benefited the commune. A hobby-farm tax exemption."

Hershorn chuckled. "You are implying something unethical, I gather. Not at all. Besides, that's how most bills are made. It's who you know."

"Did Gordon contribute money to your campaign?" Drayco spotted other photos he'd missed earlier on the man's desk. Corporate-looking types. Or lobbyists.

"Contributors are public record."

"I checked. But this one was likely off the books."

Hershorn shifted on his feet. "Gordon did give some money, yes."

"Where did it come from?"

"I assumed it was his savings."

Drayco picked up a small bust of Abraham "Honest Abe" Lincoln. "Oh? Then why the secrecy?"

"It was to be. . .anonymous. Nothing illegal. Just that whole vow of poverty thing. He probably gives to a lot of charities that way. Who am I to stop him?"

"Did anyone else from the commune pressure you to sponsor any bills?"

"I don't appreciate the innuendo, Mr. Drayco."

"Not illegal pressure or bribery. Just attempts to influence you because of the 'who you know' part."

"I haven't kept in touch with most of them. Max is a big supporter of environmental causes due to his background. Well, his company's background of abuses."

Drayco nodded. "I heard about that."

"He extracted promises from me to make the environment a priority. Since I share his concerns, it was a no-brainer."

Drayco looked to the right of the Constitution frame. He'd missed it on his initial survey, but an aerial photo of Gaufrid Farm hung among the wall collage. "What were your impressions of the remaining commune members—Gordon, Seal, Max, and Ivon Leddon?"

"All collegial and hard-working. I was only there for a few years."

"You heard about Ivon's disappearance?"

"Gordon called to see if I'd heard from him, but alas, no. Very strange move on Ivon's part. I suppose he wanted to disappear into a new identity. Happens more than you know."

After being in the FBI for ten years followed by private consulting, Drayco did indeed know. Not that Ivon fit that profile, and Ivon's cousin certainly disagreed. "What about Lara Davidenko? What were your impressions of her?"

"Lara was. . ." He sighed. "Something of a lost soul. Like she didn't know who to be. The type that morphs like a chameleon to suit whatever the situation required. Not exactly a holistic thinker."

"Catherine used that phrase, too. 'Holistic thinker.'"

"Guess I picked it up from her or vice versa. Who knows?"

Drayco put the bust of Lincoln down gently. "I'm curious. . .did you know Lara's real identity?"

"What do you mean?"

"That she was actually raised as Minna Hallow in a small Maryland town."

"Are you serious? But that accent, that whole sob story. . ."

"An act, basically."

Hershorn crossed his arms over his chest. "I truly had no idea. She missed her calling. Should have been an actress. And here I was feeling sorry for her."

"Did you ever see her abuse drugs?"

"Drugs? Not openly, no."

"Any drug use among the other commune members?"

Hershorn got defensive again. "Like me, you mean?"

Drayco smiled briefly. "My only interest is justice for Lara. Since you have an airtight alibi, you should feel free to answer questions and assist law enforcement, am I correct?"

"I will help as I am able."

"I ask because of Niles Peto, who died of an overdose. And Daven Monk, who was also a cocaine user."

"Cocaine? I knew about Peto, but Monk hid that pretty well. I'm beginning to think more went on behind the doors of those cabins than I suspected. Maybe that's why Monk went into the monastery."

Hershorn picked up a pen that he tapped idly on the desk. "That was a real shocker. The commune's one thing, but a monastery?"

"As you say, penance. Was there any ill will between the victim and the other commune members? Any sexual dalliances? Blackmail?"

"You don't pull any punches, do you? You take after your father, and I mean that as a compliment. I think." The representative smiled a bit more broadly this time. "The sex, sure. But blackmail? I didn't see that."

"There is evidence Lara was blackmailing members of the commune. I have to ask if she or anyone else tried to blackmail you because of Catherine? Or used her to get to you in some way?"

"I see where you're going with this. Lara never attempted to blackmail me, and I certainly wasn't aware of the others."

Hershorn looked at the clock on the wall. "I'd love to reminisce some more," although his expression indicated he'd rather not, "But I have a meeting in fifteen. It'll take that long to walk over there."

Drayco leaned over the desk to shake Hershorn's hand again. "I really do appreciate your time. If you think of anything else. . ."

"I'll give you a call. On one condition."

Drayco's eyes widened. "And that is?"

"I gather you'll be over at the farm during your investigation. Keep an eye on Catherine for me, won't you?"

With a promise he'd do that, Drayco exited the Longworth House Office Building and started his own thirty-minute walk back to his townhome. All that walking gave him time to go over his meeting with the representative. Most politicians quickly learned the fine art of schmoozing and pretending to cooperate with others. More often than not, they were guilty of what Hershorn had accused Lara of, morphing chameleon-style to suit whatever the situation required.

Was Hershorn honest, or had he learned to weaponize half-truths and lies? Politicians were a lot like fencers—with the fencers, it was parrying and thrusting, while with politicians, it was avoiding and evading. Effective politicians and fencers were equally good at hiding their moves and their positions.

Either Lara had uncovered blackmail fodder about Hershorn, or she tried and failed. Yet, the representative *had* hidden his relationship to Catherine from Aronson. And he took pains to cloak Aronson's donation to his campaign that resulted in a tax break for the farm.

Drayco had another sudden wave of sympathy for Lara. He wasn't about to lose sight of his primary quest to free a potentially innocent man. But he was growing equally determined to find justice for little Minna Hallow, the "lost soul," as Hershorn described her.

She was definitely a victim, of murder if not more. It was also possible she was merely a bit player in the scams around her, manipulated by others. After all, the deadly dance of a puppet always ended when the puppeteer cut the strings.

That was the part of his job that drove him the hardest—to untangle the threads of those who'd been manipulated, to recover the voices of the dead, to tell the world there was something in a lost life that mattered. Justice could be blind, it could be deaf, it could be fleeting, but if Drayco had anything to do with it, it would never be mute.

29

Benny Baskin had set up another jail interview with Harry Dickerman, and Drayco was only a mile away from the facility when his cellphone rang. He answered it, thinking it was Baskin, but instead, he heard, "Did you rat me out? After I asked you not to, and you said you wouldn't?" Alisa Saber's voice was filled with a shaking anger he could hear over the phone.

"What happened?"

"The police in Fairfax. They called me in for questioning about me playing a reporter and interviewing Harry. They know about Tasha, too. They're going to charge me, I just know it. I don't have a lawyer, and I didn't know what to do. So I called you."

"I'm on my way to the Fairfax detention center right now to interview Harry. Don't volunteer any information to the police, okay?"

"What do I tell them?"

"Tell them you're exercising your right to an attorney."

"But, as I said, I don't—"

"Just tell them. I'll be there soon."

Drayco was now only five minutes out from the detention center, pulling up right as Benny did. When he told the attorney about Alisa's predicament, Benny hurried inside and told the detectives he was her attorney, and she was not to answer any further questions without him. He also demanded to know if they were charging his "client."

When they ushered Benny and Drayco in to see Alisa, she was sitting slumped in her chair, looking more like five than twenty-five. She lifted her head and aimed a scowl at Drayco. "I have to know—are you the one who told them about me?"

"No, I didn't. I told Benny Baskin here, but he wouldn't have alerted the police. Did *you* discuss this with anyone other than me?"

She fixed her gaze on the table for a moment before replying slowly, "I might have mentioned it to a couple of friends at school. I guess they could have told somebody else. Especially Gina. I've seen her get pretty sloshed at bars."

"This Gina. . .do you know her well?"

"She's not a close friend, but. . ." Alisa groaned. "Her cousin's a cop. Oh god, they think I did it, don't they? Murdered that woman because of what? Jealousy?"

Benny piped up, "Detective King says they're only considering charging you as an accessory. If the evidence supports it. Which it doesn't."

"But I have an alibi."

Benny grimaced. "They talked with your professor. He only keeps loose attendance records. Something about an honor system, bah. Not like the strict classroom discipline of my day."

She blinked back a few angry tears. "That's great, that's just great. And everyone told me to take Professor Heyer's class because he was easy."

Benny patted her shoulder. "Don't you worry. I'll have you out in two shakes of lamb's tail."

Her look of confusion almost made Drayco laugh. He interpreted, "Mr. Baskin means you'll be walking out of here shortly. And one of us can drive you home."

She nodded, and her shoulders relaxed a fraction. "You said you're going to talk to Harry Dickerman, too?"

Drayco replied, "Right after we get you out, yes."

With a wistful look, she said, "Wish I could talk to him."

Drayco exchanged a look with Benny, both of them remembering Harry's statement that he didn't want to talk to her—or for her to see him in jail. Maybe he'd feel differently now? It might not hurt to ask him one more time.

Once Alisa was released, and they'd decided to call a taxi for her, Benny and Drayco returned to the official visitor rooms to wait for

Harry. But any hope he'd want to meet his potential daughter after hearing of her plight quickly flew out the window.

After sitting there stunned at the news, Harry said, "I want you off the case, Mr. Drayco. And Mr. Baskin, I'm confessing to everything. I killed my ex-wife, Lara Davidenko. It's all on me. It was premeditated, and that's the end of that."

Drayco sat down across from him. "You know we see right through you, don't you? You're trying to protect someone who could be your daughter. If you do cut us loose, it will be harder to uncover the truth. Or to get justice for both Lara and Alisa."

Benny added, "I'm representing Alisa, should it come to any charges. We've already got her safely on her way back home as we speak."

"Okay. But I'm only giving you a few days to sort this all out. Because if Alisa is charged, then I'm going to confess to the murder. To that Detective King."

Drayco leaned forward on the table. "I'm surprised you're willing to go to such extremes for a daughter you've never met. And refuse to talk to."

Harry wiped his brow with a hand that was shaking. "I have my reasons. And Alisa is innocent. She has to be."

He didn't look as convinced as his words were. But still, Drayco was impressed by Harry's gesture. He was even grudgingly beginning to have some respect for the guy.

Benny said soothingly, "We'll take care of Alisa. I promise you."

"I'm holding you to that." Harry glanced at Drayco with his lower lip jutted out. "Darcie came to visit me a couple of days ago. Said the two of you had dinner together."

"If you're wondering about anything improper, you shouldn't be. She told *you* about it, right?"

Harry blinked his eyelids slowly. "And she did say it was just to talk about the case."

Drayco winced at that. It was mostly true, but Harry didn't have to know about the more frank discussion he and Darcie shared. "I was on the Eastern Shore looking into the commune. So I stopped by Cypress Manor to update Darcie, since she's our client."

"Okay. I guess."

Drayco was happy to turn the topic back to the case. "Speaking of the commune, did Lara or Stuart Wissler try to blackmail you?"

"What?" Harry blinked at Drayco. "Blackmail?"

"Lara was apparently an expert at it."

Harry shook his head. "Never. Just took my money and ran."

"Have you ever worked with these companies?" Drayco rattled off the names of the businesses where the various Gaufrid Farm members were once employed.

"Not a one. But you think this is some corporate sabotage thing?"

"We're not sure yet. But since they're not ad agencies, it's a remote chance. I needed to verify a hunch."

To Drayco's surprise, Harry pressed a buzzer on the table to signal the guard to return early. "I'm tired, gentlemen. I've told you all I can." As the guard came to lead Harry away, Harry said over his shoulder to the other two men, "Remember what I said. You have a few days."

Drayco and Benny made their way back to the front lobby, and after they'd exited the building, Benny grumbled, "That was a big nothingburger. With sour sauce on top."

"Maybe Harry's feeling the isolation and stress of being in a six-by-eight jail cell. He's used to much more luxurious accommodations."

"Yep, no pools and giant fountains in there."

Drayco looked askance at Benny as they headed to the parking lot. "You're representing Alisa even though there's no legal agreement yet. Is that ethical?"

Benny waved his hand. "Ethical shmethical. We're good."

They reached the Starfire, and Drayco leaned against it. "Still, thanks for stepping in to help Alisa."

"If we're lucky, I won't be needed."

Drayco rubbed at a smudge on the car's door. "It was good to see Nelia Tyler the other day. But how's she holding up? For real?"

"Don't tell her, but I've gone a bit easy on the work. Just enough to give her experience. But I sneak in a thing or two."

"Like what?"

"Told her I'd won a coupon for a massage at Caelesti, but I hated massages. Too futzy."

"So you handed said coupon—actually paid by you—over to her?"

"Naturally. My wife's idea. Nelia's tougher mentally, hell, physically tougher, than a lot of men I know. Doesn't mean she's a robot."

"Seems like everyone wants a piece of her. . . her husband, Sheriff Sailor, the law school, even you. She's pulled in all directions."

Benny exhaled loudly. "More like drawn and quartered. I think she'll make it long enough to get her J.D. But at what cost?"

The thought of Nelia suffering a nervous breakdown seemed unimaginable to Drayco. Yet, like Benny said, she wasn't a robot. Benny was also correct that Nelia was tough, and she'd *probably* be fine. But if not, if something finally did break her, then heaven help the universe. And heaven help him.

30

Another thunderstorm, another thunderous piece. Drayco started playing Liszt's "Orage" from *Années de pèlerinage*, but switched to the same Beethoven sonata he'd played during the storm a couple of weeks ago. Or as much as his arm would allow without cramping.

As he reached the climax of the furious cadenza, a sound that wasn't thunder intruded on the music. Realizing it was a knocking at the front door, he got up to see who'd be foolish enough to come over at this late hour in the pouring rain. Surely not Darcie again.

It wasn't Darcie. He stared at Nelia Tyler standing on the doorstep, dripping from head to toe. "Good God, Nelia, you're drenched. Come in. If this is for Benny, couldn't it wait?" Especially after everything the attorney had told him about going "easy" on Nelia.

She stepped inside, her dripping clothes making puddles on the floor. "I stood outside your place for several minutes. Arguing with myself about whether I should knock or just turn around."

Her eyes were red, her face was pale, and she was shivering, although he didn't think it was from the chilly rain. "What in the world—"

"Tim's having an affair." Her words came in a rush. "With Melanie, the aide we hired to help him. Mom and Dad are furious, and Dad wants to kill him. Maybe I even want to kill him. But I'm still reeling from my parents' own impending divorce, and I don't know what to do."

He said, "Come with me."

She meekly followed, and he led her to the guest bedroom and its bath. "I'll tell you what you are going to do. You are going to get out of those wet things and take a hot shower. I'll dry your clothes for you, and they should be ready in a half-hour. There are some towels in there, and I'll scrounge up something for you to wear. I'll lay it out on the bed for you."

He closed the bedroom door and listened until he heard the shower going before he tentatively peeked inside and saw her wet clothes on the floor. He gathered them up and headed back downstairs to the laundry. Was she allergic to dryer sheets? Should he wash the clothes first? Deciding that drying "as is" would have to do, he popped them in and grabbed a pair of his sweatpants and a shirt and robe for her, like he'd done for Darcie.

He leaned against the counter in his kitchen, fighting the urge to join Nelia's father in some of that Tim-killing. After all she'd sacrificed for that jerk, working to help him through law school. Then putting her own law career aside, because they couldn't afford it, and taking a deputy job instead. Not that this came out of nowhere. Any man who'd lift a hand to his wife, as Tim had on a couple of occasions, wouldn't think twice about cheating.

Best not to dwell on it, or Drayco would call up the guy and give him a piece of his mind. So he emptied the dishwasher instead, slapping the plates on the counter harder than he should, and scrounged around to find ingredients for a hot toddy. Not that it would be nearly as good as anything Maida Jepson could make. But hopefully good enough.

He also remembered he hadn't checked the stray cat's food dish outside under the little shelter he'd made for her, and took care of that and the water bowl while he was at it. When he came back inside the kitchen, Nelia stood there in his robe, rubbing her hair with a towel. She looked at the bag in his hands and read the label. "Cat food? I didn't know you had a cat."

"I don't. Well, not officially. It's a feral cat that's been hanging around outside. I fed her once with some table scraps, and she kept coming back. She won't come in the house full-time. Pops inside, looks around, and then leaves."

"Maybe she would like to come out of the rain, do you think?"

"There's the shelter I made. And I think she has some other hiding places I don't know about. I've trained Brock to feed her when I'm away. He even tried to play with her once."

"Now that would be something to get on video," Nelia said with a slight smile.

"I wish. Got to find a way to do it surreptitiously."

"Still calling your father Brock?"

"Did I? Force of habit. Guess I got used to it all those years. But I'm working on it."

"What's the cat's name?"

"I'm not sure. She answers to 'Cat' for now, but I guess that won't do. She does seem to like piano music, so there's that."

"I like piano music too. Always have. Could you play something for me?" She looked unusually vulnerable standing there, like she might topple over.

He replied, "Sure thing. But first, a drink to warm you up." He mixed the ingredients he'd snagged from his cabinets with the liquid warming on the stove and handed it over.

She took a sip. "That's pretty good. What's in it?"

"A pinch of whiskey, honey, lemon, and cinnamon. Maida has inspired me."

That made Nelia smile a little more. "She's quite good at inspiring people. Via her personality and her cooking." She headed for the piano in his living room and turned her head to make sure he was following.

After she sat on a chair nearby, she waited expectantly. But what to play for a woman who just found out her husband was cheating on her? Some Beethoven, perhaps? Or maybe. . .

He eased into the opening notes of a piece that was familiar to him, if not anyone else. Before long, he was lost in the music and played all the way through before remembering he had an audience. He turned to Nelia apologetically but was surprised to see she was wiping away a few tears.

She asked, "What was that? It's so hauntingly beautiful."

He rubbed the back of his neck. "It's ah. . .it's something I wrote."

She stared at him. "You composed that piece?"

"Yep, a Drayco original. Something I've been toying with. Writing my own pieces, kind of as psychic therapy. I've written one after every difficult case lately."

"Even your mother's?"

"Even that. In fact, that was the piece I played for you."

She eased out of her chair to stand next to the piano. "That was absolutely wonderful. What are you going to call it?"

"Guess I haven't decided a name for that, either." He studied her still-pale face and watery eyes. "Are you okay? Can I get you something else—"

"No, I really should—"

"You don't look safe for driving yet."

He jumped up as she started swaying on her feet, but when he reached out to steady her, she fell forward against his chest. And that's when he realized she was wearing his robe without the sweatpants—or anything else. She seemed so shattered and fragile, he didn't know what to do.

As she looked into his eyes, he saw more than just emotional pain and grief. The intense longing and desire were almost too much to resist, and he drew back. But she grabbed onto him, pulled him close, and cupped her hands around his head to pull him down into a deep kiss. An amazingly deep kiss.

Afraid of taking advantage of the situation, he again tried to pull away, but when she started unbuttoning his shirt, he was truly lost. All the alarm bells going off in his brain and all the looks of disapproval he imagined from friends melted away as she clutched his hand and led him toward his bedroom. It was fierce, raw, passionate lovemaking, no holds barred, no thoughts of anything or anyone else, as the universe fell away around them.

ↀ ↀ ↀ

When he awoke the next morning with bright sunshine streaming into the room, the events of the night before hit him full force. He turned to look at the bed beside him, but it was empty. There were no signs that anyone had been there at all.

Making his way downstairs, he looked around, but no Nelia. He checked the laundry area, but the clothes he thought he'd put in the dryer for her were gone, and his robe lay in the laundry basket.

He climbed up the stairs again, two at a time, but when he inspected the guest bedroom, no traces remained of her purse or the items he'd rescued from her pants pocket when he did her laundry. Or thought he'd done her laundry. Was he going nuts? Another of his hypnagogic dreams? No, those were nightmares, and this was most decidedly more pleasant.

He sat on the bed and rubbed his hand through his hair. The overhead light glinting off an object on the floor caught his eye, and he bent down to pick it up. It was a silver feather-shaped earring, and he was pretty sure it wasn't there before. He'd never seen Darcie wearing anything like that. But he'd pulled something similar out of Nelia's sopping-wet pants pockets before taking them to laundry. Did it fall off the bed?

After throwing on some chinos, he returned to the kitchen to make a pot of coffee and continued stewing over Nelia's visit and hasty departure. When had she left? Right after he fell asleep or this morning?

He stood there for several moments, looking absently at the coffee mug, and then finally poured himself some of the hot liquid. But first things first, he had to grab some crunchies for Cat's bowl. He opened the backdoor and looked around, but no signs of Cat, either.

He drained his coffee and didn't bother putting salt in it this time, followed by a quick shower. While toweling off, his gaze landed on the necklace on the nightstand he always wore but took off last night. He picked it up as it dangled from his hand. The associate concertmaster of an orchestra in London, who'd "initiated" him into the world of sex at sixteen, gave him that Celtic dragon. When he told Nelia that story a few years ago, she'd first said the violinist could have been arrested for statutory rape—until he assured her he was six-three by then, a baritone, and a willing participant.

But she'd also wondered why he still wore it after all these years. What was it he'd told her? That it was a reminder, not of the violinist per se, but for the kindness she'd shown him, all alone in a foreign city over Christmas. Sex could be a comfort, a bonding, or a weapon. Which of those was it last night?

From the moment he'd met Nelia, there was a connection, and he couldn't deny he'd wondered what it would be like to make love to her. Just not under these circumstances. But, since she'd chosen to leave while he was sleeping rather than confront what had happened between them, he had a sinking feeling she was regretting it. Was *he*? He wasn't a one-night-stand kind of guy, and certainly not with Nelia. So, where did that leave them? Should he call? Should he not?

Women were always leaving him, it seemed. His mother. His fiancée. Darcie. Nelia. Even the cat.

Drayco sighed and slipped the necklace over his head. Why did he have the feeling he'd soon be making an appointment with his onetime therapist, Dr. Franklin Kinder?

31

Tuesday, September 29

Still half-wondering if last night was a dream, Drayco needed a distraction, and he knew where to find it. Thanks to an invitation from Sarg, he headed down I-95 toward Quantico, the BAU offices, to be exact.

He stopped by the front desk to pick up his guest pass and wait for his official "escort." This was only the third time he'd been here since he left ten years ago, and this was the first time he felt a little less awkward.

Sarg strode into the lobby and made a beeline for Drayco. "Would have headed up your way, just for a break. But I'm in the middle of that serial assault case. Can't get to D.C. because of the damned traffic. I-95 is worse than a parking lot. Hell, the train's faster."

"Onweller still steamed about you flying with me to Maryland?"

"And getting in the middle of *your* case—and local PD—without it being official and all? Yep."

Drayco spied a man entering the building wearing a straw trilby hat. Speak of the devil. BAU Chief Jerry Onweller was heading toward the elevator when he noticed Sarg and his guest. Onweller stopped, pivoted, and made a beeline for Drayco, who steeled himself for a veiled diss or one of his former boss's growling sneers. But the man shocked him.

"Scott Drayco, surprised to see you here. Changed your mind about returning to the Bureau?"

Drayco replied, "Actually, I'm visiting Sarg."

Onweller's face fell. "I hope Sarg relayed my offer about bringing you back into the fold. Could even learn to be less of a hardass if it'll help sweeten the deal."

"I'm flattered by the offer. And I've learned to never say never."

"That's the spirit." Onweller reached out to shake his hand. "You'll know where to find me. Same office."

After he'd left, Drayco said to Sarg, "Does he still have the signed photo of J. Edgar hanging on his wall? And that same god-awful cherry room freshener?"

"Yes and yes." Sarg cocked his head. "If I got him to switch to a salted coffee aroma, would that lure you back?"

"You, too? You'd really want me here?"

"I'd be lying if I said no. But I fear you'd be bored out of your pointy gourd."

Drayco gave him an imitation of a smile, but smiling wasn't easy today.

Sarg studied his face. "Uh oh. Haven't seen that expression since. . .what was her name, Eva? You know, junior, Elaine and I had been married ten years when I was your age."

"It's not woman trouble." Okay, maybe it was, but that was wholly unfair to Nelia. He'd spent the entire morning wondering if he should try to call her. Picking up the phone, putting it back down again.

He deflected with, "I was thinking about Alisa and her journalism subterfuge. Asking myself if I've totally missed the mark on judging her character."

"You, a poor judge of character? Was always one of your strong suits. I could've pointed you at people like a human-character-radar-detector, you were that good."

"But not infallible."

"No one is." Sarg pushed the button on the elevator, and they navigated toward his office.

Drayco pointed to a large round object hanging on the wall. "This serial sexual assault case of yours must be rough. The dartboard's back." He grabbed a couple of darts from Sarg's desk, judged his aim, and then threw one at the board. It went wide into the out-of-play zone. The second one, however, was a bull's-eye.

He sank into a green chair he hadn't seen before. "You said you had something to show me about Tasha?"

"Yeah. Thought it would be easier to do via my secure computer than over the phone."

Drayco scooted his chair closer to Sarg's desk to see the monitor better. "What do you have?"

"Not as much as you'd like. Checked VICAP, NCIC, Interpol's CAF, the National Sex Offender Registry. No criminal dirt. But this Stuart Wissler guy did have a passport he used to fly to Russia. On more than one occasion. This is going back thirty years."

"Sex trafficking?"

"I doubt he was collecting Russian stacking dolls. One of my Russki counterparts knew about the guy. Wissler was closely monitored whenever he was over there. Mostly known as a wannabe trafficker, but didn't have much luck. Guess he gave off stay-away-from-me cooties."

"Except to Tasha."

"A rare 'success' story of his. It's as close to proof of his international crime attempts as we get." Sarg tapped on his computer monitor. "And then, there's this. The main reason I didn't want to discuss this over the phone."

Drayco squinted at the screen, but it was blurry. Maybe it really was time he got some reading glasses.

But when Sarg saw what Drayco was doing, he banged on the monitor, and the display was suddenly clear. "Sorry about that. Budget cuts. This thing is still one pay cycle away from the junk heap. Keeps going in and out on me."

Drayco read the details. "A Justice Department arrest report?" But it wasn't about Stuart Wissler that he could tell.

"You don't see Wissler's name, yeah, I know. But this details an arrest for Congressman Curt Goldfeder."

Drayco frowned. "Goldfeder. That's the same guy who had a party where Harry Dickerman met Tasha, fell in love, and had sex with her, or so Harry said. And when he woke up, she'd vanished." So he and Harry had something else in common other than Darcie.

"You've got that look again."

"What look?"

Sarg wisely changed the subject back to the record on the screen. "Anyway, this congressman was busted for having sex with underage prostitutes at a party. Only those minors weren't all-American."

"So, trafficked."

"And this is where Stuart Wissler comes in. Doesn't say on this report, but I found a former associate of Goldfeder, who said Goldfeder later blamed Wissler for setting him up. Said it was all a mistake, and he didn't know the girls weren't of legal age."

Drayco reached for another dart and threw it with more force than he'd intended. "Just like Seal Hettrick. Wish I had a dollar for every time I've heard that."

"Would explain how Wissler dragged Tasha Oleneva into that circle."

"Probably went to Russia posing as a talent scout looking to get women into modeling or acting. But in reality, he ended up being a pimp. The women were dependent upon him because they didn't have money to leave."

"Or were afraid of what he'd do if they tried to go back."

Drayco nodded. "Gotta wonder if other commune members were Stu Wissler's partner in the wannabe-trafficking-business."

Sarg tapped on his keyboard to call up another file on the screen. "Oh, and this guy was also at that same party with Goldfeder. Busted for the same thing."

Drayco read the name on the monitor. "Henry Ofner. Sounds familiar. Wasn't he a congressman at one time?"

"From Alabama."

Drayco checked the date on the DOJ record. "Twenty-four years ago. Doesn't match the date of the party Henry Dickerman attended. But I suspect Goldfeder had a lot of those parties."

Drayco thought about Representative Bryce Hershorn, but he only became a congressman two years ago. Didn't seem likely he'd be involved in those types of parties. Still, it was worth following up. "Did you find anything overseas about our missing man, Ivon Leddon?"

"Not under that name. Could have got a passport under an alias. Like Lara Davidenko did."

"Have to wonder about Lara continuing to stay at the commune after Wissler died. It's been eight years, and yet she hung around."

"Perhaps she had nowhere else to go by then."

Drayco put his feet up on Sarg's desk and was rewarded by a shake of the head from Sarg, who hated it when Drayco did that. When Sarg did it, it was fine.

Drayco said, "We now have our answer about how Lara chose Harry as her patsy. Wissler had to know about Harry's hookup with Tasha. Likely even suspected Tasha's baby was Harry's. Then preyed on his memories of his lost Russian love with Lara's fake sob story."

"Pretty sick."

"It is that. By the way, get anything about that name on the adoption worker's paper, Gerasim?"

"Not so far." Sarg swiveled in his chair to the right, then left, a habit of his. "I've decided to give you a gold star for taking on Harry Dickerman's case. You not liking the guy and all."

"I didn't say I didn't like him. I was just wary. Rich, older man snagging a much younger wife. It's often a predatory thing."

"You still think that about Dickerman?"

"He offered himself up as sacrifice when a daughter he's never met was in danger of being charged. A narcissist or sociopath wouldn't do that."

"True. They can be good actors, but not that good."

Drayco dropped his feet to the ground and grabbed another dart. This time, he wasn't as vicious, and the dart gently hit the center with a satisfying "thunk." He asked, "Did you get that other name I sent you? Anikey Petrov?"

"One of your alleged office goons. You sure it's the same guy?"

"He's the only Anikey with a Vienna, Virginia address. Did you find anything on the guy in the files?"

"No rap sheet, *if* it's the same guy. No connections with Stuart Wissler or Tasha Oleneva or Lara Davidenko. Despite the Russian name."

"Do you have a photo?"

Sarg tapped on his computer keyboard and called up a different screen. Drayco studied it for a moment. "Looks like the same guy. I told you about the tattoo, didn't I?"

"Sounds like an *otritsaly* tattoo."

"That's what I thought. And he might have been a Russian prisoner at one time. But the hat-wearing-cat tattoo can also signify someone who's cunning and deceptive. It's not always a prison tattoo."

Sarg swiveled in his chair again. "I'll dig a little deeper with my counterparts across The Pond."

"And you didn't find any goon connections to Harry or any of the commune members?"

"There's scant info on this Anikey Petrov guy. Legal visa. And not in any crime databases."

"Yet, he just happened to choose my office to burgle out of all the offices in the greater Washington, D.C. area."

Sarg's laugh was more like a bray. "If those two burglar-goons are ordinary thieves, I'll eat my old Army boots."

That made Drayco grin. "With hollandaise or without?"

"Hollandaise on boots? Never tried that one. Although Elaine really likes my hollandaise."

"Your wife is lucky to have a talented gourmet chef at her beck and call."

"She just likes not having to cook." Sarg got up to dig the darts out of the board. "Wish you could stay and chat, but I've got an Onweller grilling in thirty."

"You owe me lunch, then."

"Deal."

"Where are you off to next?" Sarg tossed the darts on his desk.

"A monastery."

Sarg did a double-take. "Even if you're having woman trouble, isn't that a bit extreme?"

"A former commune member, Daven Monk, moved there. And yes, he's really a monk. I have some questions for him."

"Ah. Well, too bad you aren't going to the Eastern Shore because it would be tempting to tag along despite my workload. Been wanting to meet Sheriff Sailor."

Drayco grimaced. "Not in this lifetime. With the two of you ganging up on me, no way."

Sarg grinned. "Then I'll make sure I do."

32

This time, Drayco had to search for Daven Monk, who wasn't in his normal cheesemaking zone. Drayco felt a little like a hunter, trying this path and then the next, until he finally spied his prey, bending over something on the ground.

Drayco approached him and realized the man was humming a tune. "Are you a fan of Schubert?"

Monk stood up so fast, he started to topple over before righting himself just in time. "What?" He frowned at first, but his face quickly relaxed when he realized who had snuck up on him. "Guess I was humming 'Du bist die Ruh.' One of my favorites. We're not the type of order that sings Liturgy of the Hours all day. But almost wish we were."

Looking down at the patch Monk had rooted through, Drayco asked, "I didn't know you could make cheese from those."

"Not the base, no. But you get tired of the same old mozzarella or Swiss. So I like to spice it up with some herbs and fruit. This here is for a rosemary-infused sheep's milk cheese. I cottoned onto that thanks to Seal."

"I understand Seal Hettrick is Gaufrid Farm's chief gardener and chef."

"A good one, too. Never liked liver before that."

"Funny you should mention Seal. He's the main reason I came."

"Why? Has he also disappeared?"

"Were you expecting him to?"

Monk wiped his sweating face with his sleeve. "Guess any member was as likely as any other to vanish. Get a clean start. I considered it, myself." He reached down to grab some more herbs and toss them into a colander.

"And not due to any blackmail?"

Monk handed Drayco a shovel. "Here. Make yourself useful. I need a couple of holes dug over there. Next to the ones I've already made. Going to plant some more rosemary."

Wishing he'd worn better shoes for the task, Drayco nonetheless thrust the shovel into the ground and tossed the dirt aside onto a pile nearby. After several more shovels-full, he had a hole about as deep as the ones Monk had made.

Monk watched his efforts and said, "Maybe I'll hire you on. I could use the help. But to answer your question, Lara may have had some damning gossip on others. Can't believe it would be enough to make someone 'disappear.'"

"It's possible she had a partner. Gordon Aronson, for instance."

"Gordon? Why would he be involved?"

"He took her into the farm despite her only reference being a fake attorney and small-time grifter. Maybe that's why he really started the commune, as a way to ensure constant blackmail."

Monk grimaced. "You don't believe that, do you?" He searched Drayco's face, and his shoulders relaxed. "No, I don't think you do."

"Did you know about Seal Hettrick and his porn tapes with underage girls?"

"Tapes? Lara made tapes? That would mean she was complicit, wouldn't it?"

"She acquired them but didn't make them. Doesn't mean she wasn't involved at some point. This doesn't seem like a surprise to you."

Monk leaned back on his heels. "Once, when we'd had too much Gaufrid beer, Seal told me he'd made some mistakes in his life. Mentioned liking younger women. I pressed him on it, but he swore they were 'almost legal,' not kids."

"The law doesn't usually let men off for the 'almost legal' excuse." Drayco finished with the first hole he'd made and started on a new one.

"Some girls look like they're in college and act like it. And then they don't tell the guy their real age. How is that fair when it's the guy who's charged?"

"You think that's what happened with Seal?"

"I don't know." Monk rubbed his back after standing up slowly. "Gotta get one of those gardening benches. I'm too old and inflexible to get down on my knees."

"Must make prayers hard."

With a laugh, Monk replied, "We're very twenty-first century. We have digital prayers, too. Complete with speakers."

"Lara had quite a stash of money in some accounts. Seems like her blackmail on you and possibly Seal and others was fairly lucrative."

"How lucrative?"

Drayco had received a tip from Benny earlier about what the Fairfax PD had discovered, thanks to Lara's box Drayco gave them. "She had about a million in various accounts. And fifty thousand in cash."

"Yikes. That's a hell of a lot of blackmail. She only got about five grand of that from me."

"Either the others were better marks, or she also targeted non-commune members. But Gordon Aronson allegedly has plenty of cash. Other members may, as well."

"Guess the core values of Gaufrid Farm were all smoke and mirrors."

"With lots of secrets to go around. For instance, were you aware Catherine and Bryce Hershorn are sister and brother?"

"For real? No, I had no idea. They hid it well. Should have guessed. Both smart, both multi-talented, both tough."

Monk stopped to mop more sweat off his face. Gardening in these temperatures in a long cassock and hood—even a white poplin one—couldn't be easy. Drayco was getting a little rank himself, just from the hole-digging.

Drayco said, "Catherine and Bryce's relationship could be blackmail fodder. And then there's Max."

Monk grunted. "Ah, Max. Such a good son he is. With his monthly visits to see his aging parents. Pile those halos on now."

"You don't approve of visiting his parents?"

Monk hesitated. "Sorry. It's just. . .I lost my parents years ago in a house fire. Anyway, I think Max's mother is in a nursing home now. Such a shame."

Drayco finished his second hole and decided that was enough without stripping down to his shorts. "Are you sure Lara wasn't blackmailing you for more than just your drug habit? I found arrest records from your teens and early twenties. It seems you got off lightly. I can't help but wonder if your father, a high-powered attorney, might have had something to do with that."

Monk hobbled over to a small cooler beside a bench and pulled out two bottles of water. After handing one over to Drayco, he dropped onto the bench. "My father was overbearing, but it wasn't unusual for his ilk. Most people say their fathers don't approve of their sons, but Dad was sure proud of me climbing up the corporate ladder. Drugs, sex, it was part of Dad's world. And so it became part of my world."

He took a few more sips and sighed. "I'm not proud of much of my past, Mr. Drayco. That's why I'm here. The commune was a failed attempt at penance. But I think I've finally found my peace. As the Buddha said, 'Peace comes from within. Do not seek it without.'"

"Buddha?"

"Wise words from another wise man."

"Are you sure about that inner peace thing? An acquaintance of mine who's also a stock market escapee said you can take the man out of Wall Street, but you can't take Wall Street out of the man."

Monk twirled the base of the water bottle around in one hand. "Do I miss the sex, the drugs, the adrenaline rush, the glitz, you mean?"

"Something like that. And the lure of easy money."

"To be honest, sometimes I do. Kind of like being a diabetic and craving chocolate fudge cake with ice cream. You can do it, but you'll pay the price. Eventually."

"Like Lara Davidenko, née Minna Hallow."

"Poor Lara. We didn't always agree on things, but she was a big help in the brewery with Max and me. She had a wicked sense of humor. A lot of Russian jokes, which is strange in retrospect. Knowing she wasn't raised over there."

The sun beating down on Drayco's ill-conceived choice of black pants and black shirt was taking its toll. Max was wiping his face continuously now, and when he apparently noticed Drayco's discomfort, he said, "I've got enough herbs. Want to see what monastery food tastes like? Some of my famous cheese?"

Deciding that wouldn't be a bad idea, Drayco followed Monk back to the main compound and into a dining hall with a modern kitchen. The offering of cheese was tastier than he'd expected, better than anything from the pricey natural food marts.

Monk handed over a small beaker of ale. "You're driving, I know. This shouldn't make you fail a breathalyzer test."

Drayco tasted it. Even better than the cheese. "I thought you were no longer in the beer business."

"That's from Gaufrid Farm. Max ships me some from time to time."

"That's awfully generous of him."

Monk took the beaker from him and added it to a sink full of dishes. He said over his shoulder, "I see Brother Thomas has been a touch slack with washing up today. Can't blame him. I'd much rather stick to cheese." He looked at a clock on the wall with a picture of the Last Supper on it. "I don't want to rush you off, but it's almost two. Time for afternoon prayers."

Armed with some cheese-to-go and frustratingly few new leads to pursue, Drayco returned to his car, serenaded by "Dona Nobis Pacem" piped over an address system throughout the compound. Guess Monk was right. Gone were the days of monks congregating in the arching chapels and cathedrals to sing the hauntingly beautiful Gregorian chants. This was religion for the tech era. TV evangelists, confessions over the phone, and now even canonical hours in bits and bytes.

33

Wednesday, September 30

Drayco awoke in a foul mood. It wasn't just because he'd taken to sleeping in his guest bedroom since he couldn't bear to use the room where he'd betrayed Nelia. Or at least, that's how the tryst still felt to him.

But he'd also had dream after dream where he failed Benny Baskin, and Harry was sentenced to life. His mother had made an appearance, too, as if trying to tell him something in code. Or perhaps it was his subconscious sending him messages that he was on the wrong track.

After spending the morning with more fruitless phone calls and research going nowhere, Drayco's mood hadn't improved. So, it shouldn't be surprising that when he tried to tackle some Rachmaninoff on the piano, his arm cramped up within minutes. And this time, it was so painful he had to take some pills and grab a heating pad.

He silently cursed his uncle for setting up that scholarship at UMD in his name, which in turn made the university pressure him to schedule a recital. How in the world was he going to get through an entire performance if he couldn't make it through five minutes on his Steinway at home?

He headed toward the back door to see if Cat had eaten any of the food he left out earlier. There seemed to be some crunchies disturbed, but was it Cat or a rat? He'd almost shut the door when a small silver fur ball squeezed through and marched into the kitchen.

"Guess you got tired of the heat out there." Should he get the critter some iced water in a bowl? He was well past testing the limits of his cat knowledge. This particular cat seemed to like music, so maybe he'd coax his arm to relax enough to play something easier, like Satie's Gymnopédie No. 1. Anything other than what he'd found himself playing earlier that morning, Debussy's *La fille aux cheveux de lin*. . .the girl with the flaxen hair.

Oh, Nelia. Maybe it would have been better if he'd merely given her some tea and sympathy that night and pushed her out the door. Better for her or better for him? He wasn't entirely sure, but he hadn't heard a word from her since. Not a good sign.

Before he started his cat-soothing music bribe, he needed to check the mailbox outside the front door. It was mostly an exercise in rescuing circulars and political propaganda, except for one smallish package. He looked at the label, but no return address.

He eyed it skeptically. It was too small to be a bomb, most likely. But it didn't have the usual type of postmark on bulk spam mailings. And he was pretty sure he hadn't ordered something lately and just forgotten about it.

After staring at it for several moments and wondering how to proceed, he got up to grab a knife and sat down on the sofa with the package in his lap. Bomb or no, he checked for wires or powder or anything suspicious. Then he carefully opened one corner of the box to try to peek inside. He couldn't see anything at first. But that tiny opening was apparently all it took for something small and black to crawl out onto his hand.

When he saw the eight legs, he slowly reached with his other hand for a pillow to push the spider off without it taking a bite. What he didn't count on was a silver fur ball hopping up on the sofa and batting the spider away with its paw. The arachnid flipped over on its back and revealed a red hourglass pattern.

Drayco called out, "Cat, no, that's a black widow."

The feline seemed fascinated with this new toy, and Drayco jumped up to grab a glass he cupped over the spider. When another black widow started to crawl out the opening, Drayco raced to the kitchen to snag a baggie and scooped the box and its latest escapee inside. After he'd carefully added the spider-under-glass to the bag with the box, he sealed it tightly. Just in time, it seemed, when several more black spiders started crawling out of the box. To be safe, Drayco placed the sealed baggie into a clear plastic box with a tight-locking lid.

Cat had stayed on the sofa, sitting and watching him with a curious expression. He smiled at her. "My hero, or should I say heroine. Thanks for the assist, Cat."

In reply, the feline started washing its paws as if saving a human from a toxic spider was all in a day's work. Drayco checked his spider-containment efforts one more time. There might be some microscopic evidence on the box to help identify the sender. Later, he'd contact the post office to try to track down the origins of the package.

He'd told Benny Baskin he would be at Baskin's office in a half-hour, but that meant he'd have time to drop off the package at his own office first. The spiders he'd take to a biologist acquaintance later, although Drayco wasn't sure what the guy could tell him that he didn't already know.

What to do with Cat? He didn't have a litter box, a fact he should rectify. When he opened the door, however, she marched right back out with a swish of her tail as if to say, "My work here is done. Later."

First a copperhead in his car, and now this. To be fair, the copperhead might just be an accidental phenomenon of nature—if a far too coincidental one. But the spiders most definitely were not. If someone wanted to deter Drayco from this case, it was an odd tactic. And if someone simply wanted him dead, there were far more effective ways to do it.

Perhaps this didn't have anything at all to do with his present case. Maybe a different one, recent or from the past. Or maybe Drayco's joke to his father about a ricin umbrella poke in Drayco's future wasn't that far off.

࿐ ࿐ ࿐

Even after dropping off the spider package at his office, Drayco was a few minutes early for his meeting with Benny. Right as he was walking up the hallway from the elevator, he set eyes on Nelia, who was just leaving Benny's office.

When she saw him, the expression on her face was about as readable as a greeting card after getting run through a paper shredder. But she certainly wasn't smiling. Had she secretly hoped she could leave before he arrived, thus avoiding him?

They stared at each other without saying anything. Not knowing what to say, Drayco finally blurted out, "Cat saved me from a black widow bite this morning."

Nelia's eyes widened. "Cat? You mean your feral cat?"

He nodded.

Nelia looked both horrified at his spider news, but at the same time relieved to avoid discussing the elephant in the room. "Then you should definitely give her a name. Something heroic."

"I've called her Ludwig a few times. Since she seems to like Beethoven."

"I didn't know you had black widow spiders." He saw the wheels turning in her head, wondering if there were spiders when she was there two nights ago.

"They came in a package in the mail. No return address."

She chewed on her lip. "A threat, do you think? A warning?"

"I'll go over the box later for clues."

"That's good. If I can help. . ." She kept shifting her feet on the floor, then rubbing her arm, then shifting her feet again. "Of course, my schedule's kind of full."

After standing there for what seemed like a year, still not addressing that elephant—which was now stomping around in the hallway—he pulled the silver earring out of his pocket and handed it to her. "You left this behind."

She took it wordlessly, and after staring at it, stuffed it into her pants pocket.

He sighed. "Nelia, we have to talk about this."

"I can't do this right now."

Drayco looked up and down the hallway. "But there's no one around."

"I mean I can't do us," she gestured between them, "right now."

He studied her face, her body language, anything for a clue as to what was going on in her mind. "I'm sorry if it wasn't right for you."

Her words came out in a rush. "You don't understand. It was perfect. Wonderful. Magical. I don't know. . .when you played your music for me, I guess I fell under its spell."

"So, it was the music that got to you and not me?"

"That's not it at all."

"Then why—"

"I can't be sure that it was *us* that was magical or whether it was the fact it wasn't Tim." She bit her lip. "I don't want you to be the rebound boy. You deserve better."

"Don't you think I should be the one to decide that?"

"I need more time. Time to figure all this out. Time to be independent from Tim, time to be Nelia and not Mrs. Anyone. Frankly, with law school and work, I don't know if I have time for anything else. Or anyone else. Please understand?"

The raw torment on her face was more than he could bear, but unlike Tim, he respected her. What had Nelia's mother said? True love finds a way, even if it has to take detours here and there. He didn't know if what he and Nelia felt for each other was true love or mutual admiration or lust or if it was all just a misaligned Cupid's arrow. He didn't want to wait to find out, but it seemed he'd have no choice.

She gave him a slight smile and said. "I *can* use a good friend."

"You know you can call me any time. Day or night."

"You might regret that."

"I doubt it."

She paused a moment before she excused herself and vanished down the stairwell so fast, anyone watching them would have suspected Drayco of sexual harassment or being a jerk. Maybe he *had* been a jerk. Taking advantage of a woman who'd just discovered the husband she'd sacrificed so much for was cheating on her. And then she'd come to Drayco for. . .what? Comfort? A friend to talk her off a cliff?

But on the other hand, she'd seemed to need the sex at the time, even pushed for it. Maybe that was the real source of that sour feeling in his stomach, that it was a one-off for her and any true relationship was doomed to fail.

He started to open the door to Benny's office, hoping he could slink in without looking like a boy sent to the principal's office. Then, he realized what Nelia had said to him in the hallway. Magical? He straightened up and turned the knob as he felt a little smile dawning on his face.

Benny Baskin apparently wasn't in the frame of mind for playing psychologist, as he hardly seemed to notice Drayco's appearance whatsoever. The first words out of his mouth were, "The Fairfax PD caught those goons from your office break-in. Wanna have some fun and go see Detective King gloat?"

⁂

Benny hovered nearby while Drayco conferred with Shephard King. The detective said right off the bat, "Based on details you gave, we think these are the guys who searched your office. And Lara Davidenko's, too. We've got 'em in a lineup and would like you to ID 'em. If you can."

King led Drayco and Baskin into a smallish room with a large one-way glass taking up most of one wall. King pressed an intercom to say, "We're ready, Zack. Send 'em in."

A parade of nine men filed into the space on the other side of the glass. They all wore similar slacks, long-sleeved shirts, and identical white caps. Drayco didn't hesitate as he immediately pointed out two among the lineup. "Numbers three and seven."

King raised an eyebrow. "You're sure?"

"How could I forget two ugly pans like those?"

King communicated with "Zack" again, "It's okay. You can recall the herd now." The detective folded his arms across his chest. "You picked out the two guys we snagged. Guess we can charge 'em now."

"Does the one with the bulbous nose and wide-set brown eyes have a cat-in-the-hat tattoo on his forearm?"

King nodded, and Benny piped up, "Tattoos. So easy to spot the crooks wearing them, isn't it?"

King bristled at that and folded his hand across his chest, exposing his own hissing-snake tattoo. Drayco directed a fake cough at Benny, and the attorney quickly added, "Not that there's anything wrong with tattoos."

The detective turned to Drayco, "We'll see if we can get 'em to talk now. They've denied everything, natch. They're Russian, like you thought. Got work visas. But we can't find any work they've done in months."

"They didn't re-up with USCIS?"

"Did file some papers about contracting work. Their previous references seem to be squeaky clean. Mostly labor outfits. Part-time stock clerks and such."

"Did you ask them about the audio bug?"

"Just like you wanted me to." King didn't roll his eyes, but he didn't have to. "Again, they pleaded ignorance. We're digging into their backgrounds, movements, whereabouts, etc."

"Did you find any—"

"Prints on the bug or in Lara's apartment? None. Wiped clean. Whoever was behind that is a pro, up to a point." King shot Drayco a quick sizing-up scrutiny. "You say these two men both came at you? At the same time?"

"More or less."

"Huh. Our gorilla with the buzz cut weighs at least two-thirty." King's expression had a glint of respect for a brief moment. Very brief. "We'll keep at 'em, grind 'em down. And we'll let Mr. Baskin here know of any developments."

Once back in the lobby, Baskin said to Drayco, "Good news, huh? Caught the goons, no connection to my client."

"We can't be sure of that yet."

"What do you mean by that crack?"

"It's still early, Benny. What if Harry was the one who hired these guys?"

"Tell me you don't really believe that." Drayco almost saw vents of steam coming out of Benny's ears.

Drayco replied, “Can’t. Not for certain. If it makes you feel any better, it seems far-fetched, way too convoluted and too risky. Harry’s an unlikely kingpin.”

“Okay, then.” Benny rubbed his eye patch. “Hate to give you the brush-off, but I gotta run. Calendar’s jam-packed today, and it’s just now going on noon. Where are you off to next?”

“Running. Although not in a way you would appreciate.”

Benny shook his head as the two men parted ways. Drayco was relieved the goons that attacked him were out of commission for now, but he was still uneasy. Maybe it was his black-widow encounter earlier. Or his unsettling encounter with Nelia. Or perhaps it was that feeling he sometimes got on a case, akin to a sense of dread. One of his woo-woo-spouting cousins once told him it was “premonition.”

But that was way beyond Drayco’s field of expertise, and his inner instinct-o-meter was nudging him about something else he couldn’t put his finger on. He was missing some part of the puzzle, some vital piece dangling beyond his reach. He hated that feeling. It was like an unresolved tritone in music, thc “dcvil’s chord,” and it grated on his nerves every bit as much.

34

Drayco was a good eight inches taller than his running companion, but he still had to maintain a fast pace to keep up with her. In addition to being a biology student, a part-time actor, and a fake journalist, Alisa Saber was also on the Georgetown cross-country team. She'd agreed to meet him again as long as she could keep practicing, studying, or training, and this seemed to be the best compromise.

As they ran past the Washington Monument—and past a pair of sunglasses-wearing men in suits he guessed were secret service—he asked, "Was the Fairfax police grilling difficult? Sorry you had to be put through that."

"It wasn't too awful. After Mr. Baskin stepped in to take care of things."

"With him in your corner, you're in good shape."

"Don't think the cops were all that serious about charging me. But I've never had any contact with police before, so who knows? I still have nightmares."

"I know a thing or two about nightmares."

She looked over at him with a small smile. "Guess I'll get some funny stories out of it to tell my children someday."

Drayco forced himself to maintain a steady breathing pace to match his stride so it would be easier to talk. "Did they ask you about your gun training during your JROTC days in high school?"

Alisa came to a screeching halt and put her hands on her hips. "What the hell is that supposed to mean? Are you accusing me of being the one who killed my father's ex-wife, Mr. Drayco? Because I once handled a gun, I'm an expert at knives, too? I thought you were on my side."

He stopped and faced her. "The Fairfax PD will find out about your high school training, eventually. It might give them a few wrong ideas about you, and it's my job to anticipate their next move."

"So they'll put me on top of the suspect list, you're saying? That's just great."

"I wouldn't say that. There are more cracks in the Harry-as-killer scenario every day."

"But that means they'd be happy to have me take his place."

"I don't think so. Besides, you're not tall enough."

She furrowed her brow. "What?"

"Let's just say it has to do with a tree, a yard, and a fence."

Alisa just stared at him for a moment, then shrugged and took a few sips from her water bottle while he did the same with his. She didn't wait for him to finish before she took off running again, so he had to sprint a little extra to catch up. They headed over toward the Tidal Basin and the paddle boat dock, curving around the walk among the still-green cherry trees. This time of year, most of the tourists had left, so few human obstacles stood in their way.

Alisa stopped suddenly once more, and he turned to see her grabbing her leg, with her lips formed into a grimace of pain. "Damn it. Not again."

"What is it?"

"Cramping. Coach said to make sure I get enough potassium to keep the electrolytes balanced. Guess I need to eat more bananas. Those sports drinks make me want to puke."

Drayco pointed out a bench in the grass under the trees. "Let's head over there if you can."

She hopped on one leg until they reached the bench. When she sat down, he knelt in front of her and started rubbing her calf muscles. "I'm also quite familiar with cramping."

"Your leg?"

"Right arm. An old injury."

For the first time, she seemed to notice the pinkish-white branching tree of scars running down his forearm into his hand. "Oh, that looks bad. How did it happen?"

"A teenager wanted my car and didn't care how he got it. An arm slammed in a car door is like a stick in the jaws of an alligator."

She sucked in air through her teeth. "How long ago?"

"It's been about seventeen years."

"And it still cramps up? What do you do for it? Maybe I can try that."

"I'm not sure it would help in your situation. I have to soak it in warm water before I play for any length of time. Otherwise, I rub it like this."

"Play? You mean piano?"

He nodded, and she said, "Just call us the charlie-horse champs. Or the twitching twins."

He smiled as he sat back on his heels. "Feel any better?"

She wriggled the leg in the air and tentatively put it down on the ground. "Much better."

"Just to be sure," and he started in on another round. "Did Benny Baskin tell you about Harry Dickerman's heroic gesture to save you?"

"What do you mean?"

"When he heard you might be charged for the murder, he wanted to tell the police it was all his doing. And then he tried to fire us."

"Oh, my god. No one said anything. I had no idea."

Drayco didn't feel any spasms remaining in her leg, so he stood up to keep from making his own leg start twitching. "To say Benny and I were shocked is an understatement. Harry hasn't even met you."

She looked thoughtful. "During my stint playing reporter, I found out a lot about Harry. It was all positive."

"No skeletons?"

"None. So I guess I'm not surprised he'd do that. Seems like an honorable man."

"Probably would have done the same for Darcie Squier if she was charged. I think he truly cares about her."

Alisa leaned forward and tested the leg a little more. "Suppose I'll have to meet her one of these days. I mean, if Harry is exonerated, she'll be my stepmother."

"You're only ten years apart."

"Ugh. Not sure how I feel about that." Alisa looked away for a moment and then turned to him with a chastising frown. "I heard you and Darcie dated. Before Harry."

"Who told you?"

"Mr. Baskin."

"I see." Drayco counted to ten to keep from turning the air blue. Not that Benny was out of line, considering he represented both Alisa and Harry. And she had the right to know of any potential conflicts of interest.

He said, "I should have mentioned it."

"It's okay. I understand. Relationships have ups and downs like a log flume ride. Wild, wet, messy. One minute calm water, the next, a heart-plunging drop into the abyss."

"You're perceptive for your young years."

Alisa grinned at him. "I got an 'A' in psychology. And I love log flume rides."

"Would it have changed your mind about hiring me? If you'd known beforehand?"

She considered that for a moment and then shook her head. "But you still should have told me."

"Guess we're even." He paused to let a noisy VIP helicopter transport, a White Top, fly by overhead and then added, "You were pretty convincing as a reporter. Maybe you should switch majors."

"If biology doesn't work out, who knows?" She eased herself off the bench and inched her way across the grass in a small circle. "I did a little investigating on *you*."

"You did?"

"Earlier this year, a woman was charged with murdering this guy. It was in the papers. But when I read about a Brock Drayco in the case, it led me to think you were involved, too. That woman was your mother, wasn't it?"

He didn't reply right away, squinting up at the sky with its first hints of sunset color, the beauty that always came before the blackness set in. "You see? You really are good at this reporter thing."

"I looked the woman up. But I couldn't find anything about her."

"She was gone for most of my life. I don't know much about her, either."

Alisa chewed on her lip. "Guess my cousin was right. That I shouldn't try uncovering the past. What if I find out my mother is a killer? What if she's sitting in some prison right now?" Alisa laughed bitterly. "That would be rich. I finally find my biological parents, and they're both convicts."

"I'm pretty sure one is innocent. And the other. . .we'll just have to see."

35

Thursday, October 1

Still not satisfied with the information—or lack thereof—he'd tried to drag out of the commune members, Drayco wolfed down coffee and a stale bagel before heading over once again to the Eastern Shore. He'd rather fly than drive, but he didn't want to impose on the Jepsons as a taxi service.

When he darted into Gordon Aronson's office at Gaufrid Farm, the man apologized for chain-yawning and explained, "Was up late last night trying to find new ways to cut costs." He uttered a mocking laugh. "And here I was so proud of the fact we were self-sustaining all these years."

"Not any longer?"

"Money from beer sales and produce from the gardens helps. But with only four people. . ." He rubbed his eyes. "We'll make it work somehow. Grow Christmas trees. Or try a new line of hand-crafted furniture. But no one has the woodworking talents Ivon did."

"Still no word from him?"

"Nothing. Every time I get a call, I think that might be him. But it's mostly bill collectors. Or solicitors."

Apparently, the cost-cutting extended to air conditioning. It was off in the building, making Drayco glad he'd worn a white, short-sleeved shirt in the warmer-than-usual temps for the first day of October.

He said, "I've dug into Lara Davidenko's background. Were you aware that was a false name?"

Gordon stared at him. "False? What the hell do you mean by that?"

"She was raised Minna Hallow in a small Maryland town."

"I don't understand. Not Russian? But her accent. . ."

"Her Russian grandparents brought her to this country when she was five."

The director's face paled. "Then that whole song-and-dance about running from a financier-abuser was also a lie?"

"It would seem so. That doesn't mean she wasn't a victim of something else, however. Or equal parts victim and perpetrator."

"I should have checked her references more. Was that Stuart Wissler fellow—"

"A fraud? Yes, very much."

Gordon sank into his desk chair and started to pick at his cuticles but stopped himself. "They say there's no fool like an old fool. But why come here of all places?"

"Perhaps to have a safe place to run her own criminal ring. She's a talented blackmailer." Drayco watched the director's face for any signs of guilt at having been a victim, himself, or a co-conspirator. But he seemed genuinely shocked.

"Blackmail. You mean the other members here?"

"A few of them."

Gordon reached across his desk to rub a wooden egg, which was hand-painted with a colorful starburst design. It looked like the one Drayco saw in Lara's Maryland apartment.

When he saw Drayco looking at it, Gordon said, "Lara made that for me. Called it a *pysanky* egg. A Russian tradition." His laugh was bitter. "Russian, right."

Drayco had looked it up after he left Lara's apartment, out of curiosity. "Actually, she used the correct terminology. Her grandmother may have taught her how to create those."

Gordon sighed. "The reasons behind this blackmail. . . Anything to draw the attention of police or attorneys to the commune?"

"I can't say for sure."

"I'd prefer those indiscretions stay anonymous, if at all possible. I wanted people who moved here to leave all that behind them."

"To find a sense of place and peace, you mean?"

"Peace. Whatever that means." Gordon picked up a pen and rolled it around in his hand. "Speaking of peace, I really don't understand why you're here. Wasn't the killer this Dickerman fellow?"

"He may be innocent. And I'm sure you'd want true justice for Lara."

"But I still don't see how we can help any further."

"That mission statement you showed me on my first visit. It included a vow of poverty. Yet, I understand you're worth quite a bit, with money you never divested, property you still own. Was Laura blackmailing you because of it?"

Gordon tossed the pen aside. "Oh, for heaven's sake. Lara wasn't blackmailing me for anything. I resent the implication I killed her over it."

"It's my job to ask hard questions—if I'm going to find that justicc."

"Fairfax police detectives were here the other day. When we told them we had no idea why Laura wanted to leave, they didn't seem interested beyond that. So why are you?"

"I have no preconceived notions about anyone's guilt. They apparently do."

"I understand. I think." He rubbed his eyes. "But it's such a tragic thing, Lara's death. And I. . .I miss her terribly."

"Were you in love with her?"

The director stared off into space. "I cared about her, I admit that. Maybe more than I should."

"Were you furious when she left the commune. And by extension, you?"

"Furious? Shocked is more like it. Disappointed. Grieving."

"What about Catherine? Did you know she carried the torch for you? And harbored some jealousy about your feelings for Lara?"

Gordon's eyes widened. "I hadn't noticed Catherine's interest. Or interpreted it as fatherly affection. I certainly never saw jealousy."

He hopped up from his chair. "But maybe I was wrong about all of that. I've been wrong about a lot of things." He opened a desk drawer to pull out a piece of paper. "I gave the Fairfax police a list of all current and recent commune members and their references. I'll give you one, too."

Drayco took the paper gratefully. "About those current members—are Catherine, Seal, and Max all here at the moment?"

Gordon reached for a ledger on the desk and flipped it open. "The car log indicates no one has left."

"I'd like to talk to them again."

The director waved his hand in the air. "Do what you need to. I want to get to the bottom of this as much as you. I only hope I don't regret it."

Drayco found Seal Hettrick in his cabin, the man's face a mask of indecision when Drayco asked to speak with him. For a minute, Drayco thought he'd refuse. But the door opened wider, and Drayco followed Seal into a space similar to Lara's and Ivon's he'd seen on his first visit. A little more modern and definitely more populated.

When the man stayed standing, Drayco took a nearby chair. "Mind if we do this seated?"

Seal nodded but continued to stand. When Drayco gazed at him expectantly, Seal finally eased himself onto a chair that looked like it was carved from a tree stump. But he perched on the edge, a skittish bird poised to take flight at the slightest sign of a threat. "I told you everything I know about Lara last time."

"I have two main reasons to be here. The first is to ask if you knew Lara Davidenko's real identity?"

He tilted his head with a frown. "I don't understand."

"Her real name was Minna Hallow, a Maryland girl."

"That's a real shocker. I had no idea. Why would she do that?"

"It's a long story I won't get into just yet. But the second reason I'm here is about you. And some tapes I found in a box of Lara's things."

"Tapes?" Seal swallowed hard.

"They were X-rated and starred you and some girls of questionable legal age."

Seal chewed on his lip and looked at the floor. Drayco thought Seal was going to clam up until he began speaking in a tone close to a whisper. "Those tapes are from decades ago. When I was in my late twenties. I swear I didn't know those girls were underage."

"Are you sure about that?"

"Maybe I had some doubts in the back of my mind. But I didn't want to believe it, I guess. A couple of them told me later."

Drayco recalled his musings the other day about his first sexual encounter at age sixteen with a violinist in her late twenties. Hell, he'd certainly looked older at the time. "And yet, you videotaped the encounters."

"Poor judgment, I admit that, but it was consensual, I swear." Drayco had to strain to hear him. "I'm not a monster. I'd never take advantage of anyone."

"You never shared the tapes with others?"

"Only the girls. They agreed to it, and we watched them together. That's it."

"And Lara managed to find those tapes?"

"Stupidly, I brought them here with me. Didn't know how to get rid of them. So I stashed them all in a locked chest in the back of the closet." He gave Drayco a rueful smile. "I tried to burn a couple, but oh, the stench. I didn't dare do that again."

"Did you ever mention any of this to Lara?"

"Absolutely not. She must have snooped around since I had a bunch of crap on top of that chest. You'd have to go searching for it."

"That's when she started blackmailing you?"

"She said she was going to send them to my daughter."

"The one who's a prosecutor?"

He nodded. "Glenda handles a lot of sex crime cases. I was terrified this would taint her work, affect her career. So I went along with Lara's demands."

"How much money are we talking about?"

"Tens of thousands. Around twenty-five grand through the years."

"Your daughter still doesn't know?"

Seal looked at Drayco with a bleak expression. "I toyed with the idea of coming clean to Glenda, but I couldn't do that to my daughter. She thinks I came here due to my wife's death and the stress."

"So you never told her the truth?"

"Couldn't bear to talk to her, knowing I was lying to her. I know she felt abandoned. And that hurts most of all."

Drayco said, "And since you arrived here? No photos or videos of underage kids to keep you entertained at night?"

Seal's face turned red. "Goddamn it, no, and I'd swear to it in any court of law on a mile-high stack of Bibles. That's one reason I joined this group. So I could get away from all of that."

He sank back against his seat. "I suspected my wife knew about the affairs. I blamed myself for her stroke, even though the docs told me it was a genetic thing. To be honest, we'd had problems before that, and if she hadn't passed away, we would have divorced. But my daughter didn't know about any of this. And I mean *any*."

Seal pleaded with Drayco, "You won't tell her, will you?"

"If it becomes necessary, I may have to."

"Then I hope you don't have to." Seal collapsed against the back of the chair and pulled out a photo from his wallet of a woman in her thirties whom Drayco presumed to be his daughter.

"So do I." Drayco told him he'd let himself out and went in search of Catherine.

She wasn't out in the "zoo" area milking cows or tending chickens this time. When he heard the faint strains of a piano, he headed for a building with a sign that read, "Amity Hall." It was a moderately sized space with an open floor plan and a large empty area in the middle, which is where he found Catherine at an older-model upright.

As he got closer, she shrieked and hopped up from the bench. "Oh, sweet Jesus, you startled me."

"I didn't mean to make you stop playing the Bach prelude."

She caught her breath and replied, "You sound like a music lover. Do you play?"

"Here and there."

She pointed toward the keyboard. "Prove it. As penance for almost giving me a heart attack, I dare you to play right now. As long as it's not chopsticks."

He sat at the Yamaha and pressed a few keys that made him wince. Not badly out of tune, but not recently tuned, either. He rubbed his right arm briefly and then launched into Debussy's Arabesque No. 1. When he finished, she said, "I've never known a Wall Street type who could play that beautifully."

"Not one?"

"What you just played was very soulful. And they sold their souls too long ago."

He swung his feet around to face her. "I was just talking to Gordon Aronson. About your interest in him."

"Interest?"

"More than interest. Something deeper."

"I was interested in him, sure. But not in love with him. I like sex, and Gordon is good at it." She fluttered her eyelashes. "I wonder if you're good at it, too?"

She flounced over to him, getting uncomfortably close, but it made him realize how tall she was, only four inches or so shorter than he was. She whirled around and made a beeline for a table in the back of the hall to manipulate the dials on a turntable. Soon, the sounds of a Strauss waltz were wafting throughout the room.

"Come," she grabbed his arms. "Dance with me."

He led her around the floor, wondering how many dances this place had seen through the years and who partnered with whom. He raised his voice to be heard over the music. "I spoke with Daven Monk. He witnessed arguments between you and Lara."

"The men argued all the time, too, but I'll bet he didn't mention that, did he? Why is it men can argue and it's fine, but when women do, it's a catfight?"

"Did it have to do with you finding out Lara's real identity?"

"Real identity? Don't tell me she was a Russian spy or something."

Drayco changed course to keep them from dancing into a mirror. "She wasn't a spy. Or really Russian. Just a small-town girl named Minna Hallow."

"Oh, that's rich. That accent, those tales of 'mother Russia' she used to tell. Boy, were we buffaloed."

"And blackmailed. That is, the other members were. Yet, as you told me the other day, you weren't. I wonder why?"

"I'm too dull."

She might be "dull," but she was a graceful dancer, and he told her so. She replied, "Believe it or not, I started out wanting to be a ballerina. Trained for years, but it's a tough life. Couldn't handle the practice."

"The arts are demanding, for sure." He eyed a table and managed to avoid it in time. "I had a nice chat with your brother up in Washington the other day."

Her pace slowed, and he almost tripped over her. She pleaded, "Don't tell Gordon that Bryce and I are related. Please don't."

"Why not?"

She pulled her arms away from Drayco and stopped dancing. "You sure know how to kill a mood, don't you? Look, I don't want anybody to put pressure on me or Gordon. For anything. Bryce and I aren't that close since we didn't grow up together. I'm the 'bastard' child and all. But he was always good to me and helped me out from time to time."

"Helped out as in money?"

She nodded. "When I left my company and joined Gaufrid, I gave all my money away as we were supposed to. But I forgot about health insurance. I had a cancer scare a couple of years ago and didn't tell anyone here. But Bryce knew and paid the doctor bills. Fortunately, it was a false alarm."

"I guess you share that with Gordon, then. A cancer scare that turned out to be benign." The music ended, and Drayco made a little bow to her.

She smiled and replied with a curtsy. "You play the piano well, *and* you're a good dancer. I imagine you have women crawling all over you."

Drayco just shook his head as images of Darcie and Nelia flitted through his mind. Not so much crawling over him as flying away as far and fast as they could. He just said, "Thanks for the dance."

"It was fun, but my break time is over. Feel free to stay and play the Yamaha as long as you like." She turned to head out the door, but before she did, she turned around. "I didn't mean to eavesdrop, but on the way over here, I heard you talking to Seal about those tapes."

"You knew about them?"

She nodded. "Seal gives money to charities that fight abused and trafficked children. Just thought you should know."

After she'd left, Drayco headed out to find Max. A delivery truck sat beside the brewery, dropping off supplies, no doubt, like the first time he'd visited Gaufrid. Although this truck had a different name, "Havisham Homebrew."

The door at the back was open wide, and Drayco saw silver-colored barrels instead of wood. He'd read about stainless steel supplies in his research, just more ways Gaufrid and the monastery weren't as old-fashioned as they first appeared. But then, manufacturers even had homebrew mail-order kits now, didn't they? Maybe he should check into getting one for himself.

Max looked up and saw Drayco as he headed over. He said something to the driver, who backed up and turned the truck around. Max watched it go and then turned to Drayco to explain, "They mixed up delivery days. We don't have room to add supplies. Storing things at the wrong time and temperature will ruin the batch."

"It all sounds too complicated to be much fun."

"On the contrary. Sure, it takes time and dedication. But the end result is its own reward."

Drayco asked Max the same question about Lara actually being Minna. He got the same reaction of disbelief that Lara had missed her calling and should have taken up acting. But Max added, "Was that why she was killed? Something to do with her fake identity?"

"We're not sure yet, but it may play a role."

Max swatted at some insects. "Should be past mosquito season. It's all the marshes around here. We once thought a member had contracted malaria."

"It wasn't Lara, was it?"

"No, it was Boyce, I think. Look, we're all still in shock about Lara's death, and we'll do what we can to help. But next time you'll call the main office? Despite what you may think, we really do a lot of work here."

Since Max seemed as antsy to get back to that work as Catherine had, Drayco decided to tour the grounds solo. After all, Gordon had told him, "Do what you need to."

He meandered toward the back of the property, where he found a well-maintained path through some switchgrass and red chokeberry. He followed it for some ten minutes to where it ended at one of the electric fences, indicating he'd reached the end of Gaufrid Farm.

Just beyond lay a flat grassy area with a wide orange ribbon tied to a square wooden fence post flapping in the breeze. A gate through the fence stood about ten yards away—an electrified gate, probably. Not that he was going to chance it.

A crabapple tree lay on the other side of the fence, but a couple of the tree's sturdy branches reached over. He'd managed it at Harry's house, hadn't he? He crouched down to get extra momentum and jumped up to grab a branch and hurdle himself over the fence.

The flat grassy area beyond was also well maintained, looking like it was mowed recently. Why here? He'd likely stumbled onto somebody else's property, but whose? He'd have to check the state records and satellite maps when he had access to his computer.

As for the orange ribbon, it dangled loosely from its post since there wasn't any wind. Property marker? Leftover survey tool? More records to investigate later.

Not seeing anything else of interest, he decided to head back. Getting over the electric fence in reverse was more straightforward, and he made it without getting shocked. As he retraced his steps, he was again struck by the area's isolation and its stark beauty. The flat farmland gave way to a spidery web of marshlands with finger-canals twisting through the saltwater web. It was easy to see why Gordon Aronson had chosen this place for his commune.

But was it all for the stated reasons of peace and brotherhood? Answering that question was possibly the key to discovering why a woman from a small town in Maryland—pretending to be a Russian pretending to be fleeing an abusive Wall Street handler—had wound up with a letter opener sticking in her back.

Maybe it was the isolation or the property or Max's mosquitoes, but he couldn't shake the feeling Gaufrid Farm was hiding something other than *gawją* and *fripuz*, neither the land of friendship nor peace. The more Drayco thought about the "farm," the more he needed a beer, himself, if not Gaufrid vintage. And he was pretty sure he knew just where to get one.

36

As he sipped from his bottle of cold beer, Drayco stared at the racks and shelves along the walls filled with lethal-looking blades, saws, and various bits and nails. "I knew it. You're a closet serial killer with the perfect cover."

Sheriff Sailor was bent over a long wooden table studded with a clamp and vises. "That's my day job. My evening job is crafting tools for other serial killers."

"What are you making?"

"This is a very scary and dangerous. . .spice rack." He held it up to show Drayco. "My wife's been bugging me for a new one."

Drayco studied the wooden object which actually had "Spice Rack" in capital letters etched into the front panel. "All these tools, lathes, and grinders, and you're using it to build a spice rack?"

Sailor glared at him. "Wait until you get married. You'll do whatever it takes to keep the peace." Sailor set the spice rack down. "You here for a few days this time?"

"I'm staying at the Lazy Crab for one night."

"Even one night'll do you some good. Put some meat on those bones."

"Why is everyone always trying to fatten me up?"

"Because most of the country is obese. You tall, un-obese types make the rest of us feel bad."

"I'll have you know I gained two pounds."

"Since when?"

"Last month."

"Probably just had a big lunch." Sailor chuckled. "You and Nelia."

Drayco was startled by that comment. *Did he know?* Then Sailor added, "I swear that woman can eat whatever she wants, and it evaporates into the ether."

With relief, Drayco replied, "You're going to miss her help, aren't you?"

"We're all going to miss her."

Drayco must have had an odd look on his face because Sailor studied him and asked, "What's up?"

"Her marriage situation. But that's confidential, so you'll have to let her decide when to talk to you."

Sailor picked up a hammer. "Think I know what it's about. Overheard Regina Reymann chatting with Nelia. Made me want to kill a certain husband."

"Get in line." Sailor was still looking at Drayco with an odd expression Drayco couldn't read, so he decided to change the subject. "Speaking of spouses. When am I going to meet Mrs. Sheriff?"

"Hopefully never because trouble follows you around."

"I said the same thing to Sarg when he said he'd like to meet you. No way I'd let the two of you gang up on me."

Sailor leaned on a ladder propped against the wall. "How is that former partner of yours?"

"Not-so-former at the moment. He's helping me with this case due to some international aspects. Since he's still official FBI, his connections can come in handy."

"Well, I can help some, too. I did some digging. Checking into a drug ring that could involve the commune."

"Find anything?"

"Drug crimes are on the rise in these parts, to be honest. But I'd never heard of any reports via the commune. Since it's a half-and-half jurisdictional situation, I had to check with my colleagues on the Maryland side up there. They said no when I asked."

"Too bad. Would help in the motives department."

"Sometimes negative answers can turn into positives."

"If they rule things out, yeah. But this doesn't mean Gaufrid members haven't smuggled drugs in and out along with the beer and broccoli they sell."

Sailor scratched the side of his neck. "Or they put cocaine in the beer."

Drayco prowled around the space to study a puzzling gizmo with a handwheel attached. "What's this?"

Sailor called out, "Don't touch that unless you really don't want to play the piano. Or don't mind missing some fingers. It's a table saw."

Drayco jerked his hand away. "Thanks. I have all the scars I need."

After he'd put lots of distance between him and the potential finger-chopping machine, he asked, "Did Daven Monk or other 'Gaufridians' come up in those PD records from your colleagues? Including the deceased member who died from an overdose, Niles Peto?"

"Seemed to keep all their noses clean. Unlike some kids who set a dumpster on fire the other day. Or the morons I had to deal with this week from Deakin's Fried Chicken who dumped used cooking oil into the landfill. As if I didn't have enough things to worry about. Don't need the Virginia EQD or the EPA breathing down my neck."

"Can't blame you." Drayco eyed the lethal-looking table saw. No way he was ever buying one of those. "Think you could run a partial plate for me? It was from a vehicle seen in these parts."

"Part of your investigation?"

"Maybe." Drayco handed over a piece of paper. "That's all I have. One letter and a vague description."

"What did this car do. . .try to run you over?" Sailor grabbed the paper.

"Truck. And run me off the road, yeah."

Sailor frowned. "When did this happen?"

Drayco hesitated. "Ah, about ten days ago."

"And you're just telling me this now?"

"It happened too fast for me to get a good look. And I did check the general databases, but nada. Thought you might have seen it driving around."

Sailor tucked the paper into the spice rack. "I'll see what I can find. Don't need rampaging trucks around here."

"Thanks."

"No prob. Oh, and FYI, a Maryland colleague did tell me one interesting tidbit about Gaufrid Farm. Happened about ten years ago. Nothing came of it, so it never made the official books."

"I'm all ears."

"Better make that 'nose.' A couple of neighbors complained about a smell coming from the place. More chemical-ish than rotting-ish. My colleagues checked it out but couldn't find anything. Put it down to marsh gases or a fish kill. During certain weather conditions, we get algae blooms. Pee-yoo."

"Has it happened again?"

"Nobody's complained since. To law enforcement. People mostly like to stay to themselves around here."

"Gaufrid Farm members, too, apparently."

Sailor grabbed a broom to sweep up some wood shavings. "Looks pretty normal from the outside."

Drayco lifted an eyebrow. "When did you see it from the outside?"

"I made a little trip up there to drive around the other day. You got me all curious."

"See anything interesting?"

"Some crop fields, a few cabins, lots of marshes. And an electric fence."

"I noticed that too. Is that common?"

"If you have prized cattle or pigs or horses you don't want stolen. Most people use the old-fashioned kind of fence. Time tested. Much cheaper."

The sheriff leaned on his broom. "Find anything to connect your client to Gaufrid? Other than the victim?"

"Not yet." Drayco took a step back and immediately regretted it when something sharp cut into his arm.

When Sailor saw the blood, he leaped into action, grabbed some peroxide and a bandage. After tending to his patient, he said, "Told you trouble follows you around. That hurt much?"

Drayco flexed his arm. It hurt, but he'd had worse. "A memento from Sheriff Sailor's Little Shop of Horrors."

Sailor pointed at the bandage. "Want me to sign it?"

"No, thanks. But in the future, let's keep our meetings to the much-safer Seafood Hut when it reopens."

37

Friday, October 2

After checking in with Benny Baskin to give him an update, Drayco drove to his destination and parked in the gravel lot, dodging hungry seagulls and blue-winged ducks used to getting handouts from humans. The white houseboat with the name *Phobos* was still moored in the same slip. But this time, Drayco noticed a new addition—a cabin cruiser was tethered next to it, the *Starfisher.*

When Drayco knocked on the door leading into the houseboat's cabin, Dennis Frischman kicked it open with his foot since his hands were full of plastic dishes. Drayco nodded toward *Starfisher.* "I see you've got a new baby."

"Needed something speedier and more nimble to get in and out of coves and bays. And still have enough storage for the day's haul."

"The marine salvage business must be doing well."

"You'd be amazed at what people will pay for some of these parts. Had a solid brass bridge telegraph go for a couple thousand at auction. A ship's anchor went for about three thousand."

Drayco whistled. "I'm definitely in the wrong line of work."

Frischman squinted at him. "I'm not usually the kind who works well with partners. But if you ever get in dire straits or need a change of pace, give me a call."

"You're the one who called me this time."

"So I did. Come on in." Frischman ducked into the cabin and deposited the dishes onto a table that he leaned against. "Found something out you might find interesting. I mean, *I* found it interesting."

Drayco claimed the same sofa bed seat as last time. "So fill me in."

"Chatted with a fisherman not too far from the commune. Lots of farmland there but also fishing spots, waterways and all. Anyway, this person has seen helicopters coming and going over the past several years. More than most farmers would need for any crop dusting or such work. He hadn't been able to find out who it is yet."

"Any pattern to the traffic?"

"No, but a lot of them are at night. Which is odd if it's farming or fishing-related. There aren't any resorts near that chopper area, either, unless you count Tangier Island. But who knows?"

"Did he get a description of these helicopters? Is it the same one, or are we talking different aircraft?"

"Since they're often at night, he didn't see much. He says they have the same engine sound, for what it's worth. I know boats, but I know zilch about planes or choppers. I think he did say of one sighting, the thing was dark with a red stripe on the side. And maybe a 'G' in the tail number."

Drayco mulled that over. It *could* be crop dusters, but at night? More like the police or military. He'd have to check with the Coast Guard air station at Elizabeth City to see if they were doing any maneuvers in that area. "I appreciate the info. I'll follow up."

"Also, after you mentioned the names of the commune members, I looked them up some more. Or should I say their former companies. Not many contenders for Upstanding Business of the Year. Or Decade. If these commune people left their businesses thinking it would somehow make a difference, they were sorely mistaken."

"You mean recent activity?"

"The defrauding, the insider trading, the environmental disasters, the 'creative accounting' still continue as near as I can tell. Stuff that hasn't been made public yet. Couldn't find any current links to your Gaufrid crew."

"Sounds like you caught some of that fish you were telling me about."

"A few teensy ones. Although you might find this morsel interesting. A source tells me Gordon Aronson is not only *not* poor, he's still actively trading."

Drayco sucked in air through his teeth. "Gaufrid Farm was supposed to be a way of escaping the rat race. . .and the rats."

"People who say they've turned their lives around are often lying. Sometimes your past just follows you."

"Like yours?"

Frischman laughed. "You looked me up, didn't you?"

"Naturally."

"And did you find anything?"

"Worked for Halsey & Hayden. One of their bigwigs, a real high roller. And then you just left. No rap sheet, no hints of scandal, no record with the FTC."

Frischman picked up a nearby antique lantern and turned it over in his hand. "I was asked to do some things I wasn't comfortable with. I went along at first. I mean, everybody did it, right? No big deal. But after a while, I realized yes, it was a big deal, and I wanted no part of it. But going to another outfit would only mean the same thing. Most people don't realize how evil Wall Street really is. It's not all ticker numbers, bulls, bears, and bell-ringing by celebrities."

"I worked at the FBI for ten years. I think I have a pretty good idea."

"I'll bet you do. A G-Man, huh?" He studied Drayco. "I can see it. So why did you leave? Having to deal with those blackguards sap your soul, too?"

It was too long a story to get into, and since Drayco had patched things up with Sarg, he didn't want to revisit that era. So he just said, "It was the right time."

Frischman put down the lantern, exchanging it for a cigarette he didn't light. "Some would say leaving a job like yours or mine is a form of betrayal. To which I say life is filled with betrayals, little ones and big ones. It's how you handle it that matters. Shows what you're made of."

"You must have made yourself quite a reputation around here. The workmen at the Boardwalk Bar seemed cowed by you."

Frischman's grin, revealing his set of perfectly matched teeth, would be positively blinding if the sun were shining on them. "Most of the traders are Ivy League boys. My past was a little more. . .unorthodox. Let's just leave it at that."

"I don't see any particular need to dig deeper into your past per se."

"I would prefer you didn't." Frischman's grin had stayed steady, but there was an added spark of warning in his eyes.

As Drayco left him, the man headed out onto the *Starfisher*, getting the boat ready for another outing into the magical world of treasure hunting—treasure in the eye of the beholder, if you were one of those marine auction bidders.

He stopped for a moment at the edge of the pier, taking in the sights and sounds. Most people would find the water lapping at the posts the most soothing part, but to him, it was the synesthesia frenzy that the water created in his brain. Sapphire-colored fuzzy bubbles. A bit like Maida's champagne Fiery Sunrise mimosa in marine form. Every once in a while, a gray-green dagger *slop!* against a hull would practically stop his heart—time to move on.

As he walked back toward his car, he studied the other boats in the slips. Most were lawful and harmless. He didn't see the boat of local thug, Caleb Quintier, but he knew the Coast Guard kept an eye out for seemingly innocuous boats that turned out to be drug runners, counterfeiters, or smugglers.

Although Drayco didn't know Frischman well, his instincts were pegging the guy as legit, and he couldn't say that about everyone he'd run into lately. Frischman may be a treasure finding master, but it was time for Drayco to do a little treasure hunting himself—of the information kind.

38

Drayco had driven by the small bus station several times on his various trips to the Cape Unity area, but he hadn't paid it much attention before now. It reminded him of tales his grandfather used to tell about traveling the country by bus, from one depot to another, from one small town to another. It was after passenger train service faded away, but before airfare became cheap enough for most people.

The station itself looked to have stayed the same for six decades, located not in Prince of Wales County nor Cape Unity, but in Temperanceville, about a mile's drive from Gaufrid Farm. It had a generic "Bus" sign hanging over the door. But when Drayco entered, it was clear the building was used as a multipurpose venue from the looks of the snack foods, soft drinks, shampoos, pain meds, and Band-Aids.

The man behind the counter looked like he dated as far back as the structure. He eyed Drayco with suspicion until Drayco casually dropped Sheriff Sailor's name. That usually did the trick. It also didn't hurt that Drayco complimented him on the fiddle-shaped antique telephone hanging on the wall.

"Got that from an estate sale," the clerk said.

"It's a beauty. My grandfather had something like that. "

"Hard to come by now. Not that anyone wants such antiques anymore. It's all video this and cellphone that. Nothing's real unless it's on one of those tiny screens."

Drayco pulled his phone out of his pocket. "Funny you should mention that. I have some folks here I'd like to ask you about. Maybe you've seen them in here buying tickets?" He called up a slide show of the photos of commune members he'd taken from his visits. Gordon,

Catherine, Seal, Max, and ones he'd made of Lara and Ivon from the picture Ivon's cousin, Porgy, gave Drayco.

The clerk studied the photos. "Think I recall this one," he pointed to Lara. "Came in here a few times. The others, I don't recall so. Not that my memory's the best these days. But I tend to remember the ladies." His lopsided grin made him look like an impish schoolboy.

Drayco flipped through the photos again until he landed on Max's. "So, you haven't seen this man here? He's a member of Gaufrid Farm down the road."

The clerk squinted at it and then put on some glasses. "Oh, yes, he comes in here about once a month. We joke all the time about him and his trips to Wilmington. Like he's gotta get his scrapple fix." He shuddered.

"Wilmington, you said? Not a Virginia destination?"

"Pretty sure it's Wilmington." After switching to Ivon's photo, the clerk studied it and said, "The commune does have an account here. But don't recall anyone like that."

Gordon Aronson had mentioned that account. . .so members could more easily travel to see family. Like Max, for instance, who was helping out his elderly parents. "Any unusual trips or requests by these individuals?"

"Well, now, that there's private information I shouldn't go into, I don't think. You can check with the director about that. His name's Gordon Aronson. Nice fellow."

Drayco thanked the man and bought some Machipongo Morning roasted coffee beans from the man's store by way of showing his gratitude.

He returned to his car, trying not to be in a foul mood prompted by news he received earlier about the goons. Since the only charges the police could bring against them were breaking and entering, they got out on bail. Somehow, he didn't think they'd abide by the standard protocol of staying "available" and checking in with the PD.

He headed for a place he wouldn't be visiting for much longer. Even now, its antebellum design with its row of gleaming white columns made it look like it belonged in a Civil War movie.

Darcie Squier was waiting for him after he'd called to see if she was there, and she grabbed his arm and pulled him into the foyer. "Look who's here."

Drayco glanced behind her and saw none other than Alisa Saber sitting on a divan in the living room. The vibes coming off both women weren't warm and fuzzy. "Should I come back another time? I don't want to interrupt anything."

"You're not," Darcie replied quickly, and Alisa had a look of relief on her face to see him.

He took a seat next to Alisa. "I didn't expect to find you here."

"I tracked Darcie down. You'd said you were going to be on the Eastern Shore and would be stopping by for a visit. Curiosity got the better of me."

Darcie joined them but didn't sit. She glared at Alisa. "I heard from Benny Baskin that he's representing you, too. You being a suspect and all. I asked Mr. Baskin why he'd do that. Isn't that a conflict of interest? I mean, why should I trust you?"

Alisa clenched her fists. "And why should I trust *you*? Why do you want to marry my father? You're only ten years older than me."

Darcie twirled a strand of hair around her finger as she stood there for a moment. But she surprised Drayco by nodding and saying, "Touché."

With that bit of detente clearing the air, Drayco said to both women, "The Fairfax police arrested two men who broke into my office. They may have been the same crooks who ransacked Lara Davidenko's apartment. If so, we have our first tangible link to a suspect who could exonerate Harry." He didn't mention the bail part.

Darcie folded her arms across her chest. "Why can't they let Harry go, then?"

"Because the men haven't been charged with Lara's murder. And Harry is still at the top of the PD's suspect list."

"But surely they see by now it's all a big mistake? Harry would never do anything like that. He's a human teddy bear."

Drayco pulled out mugshot photos of the two men he'd obtained via Benny and showed them to Darcie. "Ever seen either of these two? Around Harry, perhaps?"

"What, you mean, are they his accomplices? I thought you believed Harry was innocent."

"I *mean*, were they possibly following him, or did you see them acting suspiciously around him?"

"Oh." Darcie plopped down onto a rocking chair. "No. . .no, I'm sure I haven't."

Drayco returned the photos to his pocket and rescued another one that he handed over to Alisa. "I made an additional copy of the photo your cousin gave you. I've run it through some processing software to clean it up." He hastened to add, "But I haven't forgotten about the original photo that was stolen."

Alisa took the photo and cradled it in her hands. "Thanks."

Darcie gazed at the younger woman for a moment and then said, "I can't imagine what it's like not to have known who your mother was. And not get to be with her growing up."

Alisa replied, "Are your parents still alive?"

"Both of them. They like Harry, by the way. I introduced them about a month ago."

Funny, that. Darcie never hinted she wanted to introduce Drayco to them while the two of them were dating. But her wealthy family would be far more pleased with the even wealthier Harry than a middle-income private crime consultant. Drayco asked, "What are you going to do with Cypress Manor now?"

"It's a big, fat, steaming legal mess. Not sure I'll ever see a dime from it. What with my ex's attorneys and all. But I won't stay here, regardless. Way too many unhappy memories. I'll be glad to put it behind me."

She gave a quick look at Alisa and said, "You must stay for lunch. Both of you. I insist."

Drayco replied, "I don't have the time. But thanks for the offer."

Alisa lifted up her head. "I'll stay. If it's truly okay with you, Darcie."

"Of course." And Darcie looked like she really meant it.

As Drayco left the two women to return to his car, he realized how leaving this place felt like the end of a personal chapter in his life, no matter how Harry's case turned out. He had a few pleasant memories from an encounter or two with Darcie there, but she was also right about the unpleasant aspects surrounding Cypress Manor. Good riddance to bad karma.

He'd just climbed into his car when he got a call with a very agitated Seal Hettrick on the other end. "I got your number from Gordon. Why did you rat me out to my daughter?"

"I don't understand. What do you mean?"

"I thought our conversation was confidential. But Glenda called, demanding to know about me having sex with minors. I can't believe you'd betray a confidence like this. I'm going to file a complaint. I'll have both your head and your license for this."

Drayco tried to calm him down. "Seal, I didn't tell your daughter anything. I promise you. Did she say how she got this information?"

Drayco only heard labored breathing on the other end before the man replied, "Now that you ask, I don't think she said."

"Well, that's kind of important to know. Look, I'll see what I can find out. But in the meantime, don't discuss this with anyone else outside the three of us."

"No problem there." He added after a moment, "Gordon hasn't said anything to me. So there's that."

"If Lara were still alive, I'd expect her to be behind this."

Seal sighed. "Me, too. I've been here for a decade. Why would someone come forward now? Since the tapes were all consensual, like I told you. I'll swear in a court of law and take a dozen lie detector tests."

"I understand, Seal. And I promise I'll check this out."

Drayco hung up with an only slightly less agitated Seal and sat with the AC cranked up to full blast. He needed a recording of Beethoven's Tempest Sonata right now because it was fitting music for how the case was going. One turn, then another, then another, leading around in circles. A lot of sound and fury signifying . . . what?

He sat back, letting the cool air dry sweat from his temples, thinking about Seal's call. Who else knew about those tapes? They obviously weren't interested in blackmail, or they'd have talked to Seal before they alerted the man's daughter.

Unless it was all about misdirection, which tied into a possible motive for Lara's murder Drayco was warming up to, especially after a round of Bach fugues to jog his brain. Few things helped to focus his thoughts more than Bach's counterpoint.

He headed to the Lazy Crab for one more night before it was back to the District to give Benny Baskin an update. His client was still in jail and still threatening to confess to Lara's murder if Drayco and Benny couldn't prove Harry's—and especially Alisa Saber's—innocence. They might not have much time left if the man followed through on his threat.

39

Saturday, October 3

Drayco didn't want to put Maida to any trouble by making him breakfast before he checked out of the Lazy Crab, so he said he needed to leave early, at six. He made do with a couple of chocolate bars and a soda from a convenience store when he gassed up the car.

Four hours later, he made it to his destination, and as he entered the lobby of Miracle Hill, he immediately got the sense that the place had a wildly unlikely name. Down one of the beige hallways that branched out like spokes in a beige wheel, he spied residents sitting in wheelchairs locked in place, some mumbling to themselves, a few staring off into space. There would be few miracles for the people who wound up here.

His appointment wasn't until ten, so he stood in the waiting room studying a large display board with snapshots, drawings, and poems. A few of the folks in those snapshots, whom he assumed were more residents, were smiling. But not too many. One or two of the drawings were pretty good, but the others, scratched in crayon and markers, reminded him of the second-childhood reference from Shakespeare's "Seven Ages of Man."

The place seemed clean, yet it was hard to remove the aromas of nonenal, ammonia, and lemon air freshener that permeated such facilities. Occasionally, he heard a faint voice crying out with a repeated intonation, more from dementia than pain. The lights from an overhead fixture buzzed in rhythm with the voice, both soon muffled by a loudspeaker that began piping in a synthesized version of "Moon River."

After a few minutes, he realized he was the target of close scrutiny and spied a woman with bobbed white hair sitting in a wheelchair staring intently at him. When she saw him looking back, she waved him over.

"Geoffrey? Why, it's so long since you paid me a visit. You've grown so tall."

He started to reply politely that she had the wrong man, when she continued, "I wish you could spring me outta here. I hate that infernal music."

He replied, "Well, I'm sure they take good care of you here. And I suspect something could be arranged about the music."

The woman motioned for him to bend over closer to her, and she said, "They won't let me have a thing. You should see what I get to eat. Gray mush. Didja bring me a treat?"

Drayco patted his pants pocket and pulled out one of the chocolate bars he hadn't eaten. When he handed it over, she slipped it under a blanket in her lap and winked at him. "Our little secret," she said.

An aide headed over to their location and grabbed the chair's handlebars to wheel the woman away. The aide apologized to Drayco, "I'm sorry. I hope she wasn't bothering you, Mr. . .."

"Geoffrey," Drayco smiled. "And she was no trouble at all."

As the elderly woman winked at him again, he went back to his post by the bulletin board to wait. Precisely on the dot of ten, a woman sporting the name tag "Susan," came out to greet him and lead him down a hallway to Room 125. An even frailer woman lay propped up in her bed on three pillows watching a TV with the volume turned up to max.

Susan called out to the woman, "There's a friend of yours here to visit you."

The woman turned her head. "What's that?"

Susan called out a little louder. "A friend. Here to visit."

"That's nice, dear. Haven't had any visitors in so long."

Drayco's ears perked up at that. He thanked Susan and waited for her to leave, then moved closer to the bed. "Eleanor McCaffin?"

"Yes, that's me. Don't think I recognize you."

"My name is Scott. I wanted to ask you about your son, Max."

"Max? Is he okay?"

"Yes, he's fine. Sends his regards," Drayco lied. But he was rewarded with a small smile.

She said, "I wish he could come. He's so busy."

"I thought he came to see you each month?"

"Each month? No, I haven't seen Max in years. He's just so busy, you see."

"Who pays for your care, then?"

"I don't rightly know. I assumed it was Max."

"You must be proud of him."

"Oh, I am, I am. He's done well for himself."

Her voice was scratchy and weak, so he had to hover over the bed to hear her as she continued, "We were so poor. His father had a job that took him all over. Me or Max rarely saw him. And then he died at thirty-five, you see. Max had to take all kinds of jobs through high school. Worked his way up into a big company."

"It must have been a surprise when he went to the commune."

"The what?"

"Gaufrid Farm, the commune where he lives now."

She shook her head ever so slightly. "No, you must have the wrong Max. My Max still lives in New York. That's why he can't come see me, he's just so busy and so far, you see."

A trip from the Eastern Shore to the facility would take five hours via car, but that was hardly a world away. Mrs. McCaffin's eyelids kept falling to half-mast, and she seemed to have forgotten he was there, so he quietly headed back to the lobby and asked to speak to the director.

He was told it would be ten minutes, so he took a seat in the lobby where he was once again serenaded by the facility's sounds. The lime gravels of the buzzing fixture, the reddish blobs of murmured voices, and the squiggling amethyst chevrons of the music sent his synesthete brain into a sensory blast knitted like a wraparound Jackson Pollock painting. It was not pleasant.

Finally, a woman who introduced herself as the assistant director led him into a small office where it was considerably quieter, for which he was grateful. He said, "I'm looking into a tragic case of a victim, trying to find some closure for her family."

"I'll do what I can to help, but what does this have to do with Miracle Hill?"

"I understand Eleanor McCaffin has been here for eight years?"

"Ten, to be precise."

"I paid her a visit moments ago, and she seems comfortable and well taken care of."

"We do our best. Fighting with regulators and hunting down funding, notwithstanding."

"It must be hard. But surely it helps to have people like Max who can pay for their loved one's care directly."

She didn't seem to sense he was on a fishing expedition, so he pulled out a longer fishing pole. "I think he said he arranges it directly. I can't remember if he said direct deposit."

"Yes, it's direct deposit. Like most of our accounts."

"Right, I remember now. I think it was the Third Bank of Virginia."

"Virginia?" She frowned. "No, it's in New York."

"Of course. Where he works, that would make sense."

Drayco had looked up how much the place cost. It was one of the nicer assisted living and nursing home combos that would set you back eight to nine grand per month. Over a hundred grand per year. So, Max had money, too, like Gordon and Lara. No vow of poverty for him, either.

It seemed to be a commune epidemic—financial lies based on lies, all around. So much for utopia. At least Max was spending it on his mother. But it begged the question, what else might he be spending it on? If he wasn't using his weekend visits away from the commune to come here, then where?

As soon as Drayco finished with all the "Miracle" he needed, he ducked outside into a Muzak-free world. But his mini-reverie was soon disturbed by his Prokofiev cellphone ringtone.

It was a manager from the Havisham Homebrew trucking company returning Drayco's phone call, telling him they didn't have records of a commune delivery. Since Drayco had memorized the license and vehicle numbers, he asked about that, too, and again the answer was in the negative. No vehicles with the plate or that number.

The man apologized and said their regular bookkeeper was out of the office sick. "Maybe I don't know how to handle this computer database thingie. Give me your info again, and I'll let her know you called." He paused and then added, "This isn't about any drinking on the job or anything, is it? Our drivers are the best. Bonded and certified."

"Nothing like that."

Relief colored the man's voice as he replied, "Good."

The Havisham manager might be relieved, but Drayco sure as hell wasn't. The case was heading in a direction that wasn't at all pleasant, and he didn't like the destination at the end of that particular investigative road.

He thought back to his visit with Eleanor McCaffin. She seemed to be receiving quality care, and at those prices, she'd better be. But he hadn't noticed any flowers or plants. Not one.

He'd passed by a flower shop on his way to the nursing home, and he headed for it now to order a vase of Stargazer lilies and purple iris. Those were Darcie's favorites, so maybe Mrs. McCaffin would like them, too. What were Nelia's favorites? He'd never asked. He briefly toyed with the idea of sending her a bouquet by way of an apology but decided against it, not sure how she'd react.

He pushed those thoughts aside because he had a different woman on his mind right now. He wasn't sure how she was going to react to his visit, either. But someone had wanted to get a volatile reaction out of her, and Drayco was determined to find out why.

40

The aide showed Drayco through an office door that read "Washington County Prosecutor's Office" and in smaller letters, "Glenda Hettrick, Prosecutor," and then pointed to the black leather seat in one end of the room. The prosecutor herself was nowhere in sight, so after Drayco sat, he took the time to look around.

At first glance, it was what you'd expect of someone wanting to present a tough, professional image. Dark wood paneling matched the large desk, and several framed degrees and credentials hung on the wall. There was even a faux fireplace with a large American Eagle sculpture above it. A picture of a smiling woman in a navy pantsuit graced the wall behind the desk—all standard issue.

When no one appeared after a few minutes, he hopped up to walk around. An art glass sculpture that looked to be by Chihuly sat on a table nearby. Also arranged on the table lay a display of a dozen or so detailed model fighter planes. It would be unusual in most prosecutor's offices, but he didn't think it was a case of a woman trying to fit into the old boy's club.

After fifteen minutes of waiting, he thought he heard a female voice down the hall. He poked his head out to see a woman—in a navy pantsuit, no less—who looked like the photo in Hettrick's office, walking into a nearby courtroom. Taking a chance, he followed her and peered inside the open doorway. He was in luck because she was alone.

"Are you Glenda Hettrick?"

His question seemed to startle her, and she jumped back from the judge's bench where she was standing. "And you are?"

"Scott Drayco. We have an appointment, I believe."

She grimaced. "Do we? I'm sorry, I don't recall. It's not like Keira to schedule something and not tell me. More so on one of our few work-Saturdays. I'm not the most popular person on those days."

"I can come back some other time."

"No, it's fine, I'm free at the moment. My next case starts Monday in here," she waved her hand around the room. "Just getting my legal ducks in a row."

Drayco followed her gaze. "All courtrooms seem to have the same look anymore. Makes me long for the old-school courts with creaking floorboards and the smell of chalk dust and mimeograph fluid."

She smiled. "These new-fangled spaces offer fewer distractions, I would imagine."

"Practical, yet boring."

"What was the subject of your visit today, Mr. Drayco?"

"Your father, Seal Hettrick."

Her demeanor changed radically, and she gave a curt nod toward the hallway. "Let's go to my office."

When they were inside, and she'd made sure to close the door, she said, "Who are you, Mr. Drayco, and what's your connection to my father?"

"I work for an attorney. But one who's representing another man arrested for murder."

"Murder? My father is involved in that, too?"

Drayco replied honestly, "The police think they've got the right man in jail."

"But you don't?"

"No, but I'm not here to say your father should take his place, although he does currently reside at Gaufrid Farm. Which is where the murder victim also lived until recently."

She frowned at him and started to reply when her desk phone rang. She took the call, listened briefly, and replied with, "I'll have to ring you back later." Then she turned to Drayco and asked. "Which attorney did you say you were working for?"

"Benny Baskin."

She muttered, "Shit," but her prickly tendrils of antagonism receded a fraction. "I doubt Baskin would want to bust his perfect record. So if he thinks his client is innocent, he probably is. Meaning my father *might* take his place, after all."

Drayco pointed to the model planes behind her. "Are you a fan or a pilot?"

"Both. I don't fly those, but my grandfather did. I'd like to. But I'm more of a Cessna jockey."

"So am I." He smiled briefly. "But getting down to business, I'm here because you called your father about having sex with underage girls. Your father, in turn, called me and accused me of sharing privileged information. Was it the Fairfax police who notified you about this?"

She studied him for a moment and reached into a drawer to pull out a piece of paper she placed on the desk. Drayco scanned it. The anonymous note was computer-printed and laid out the case against Seal, making it seem far more sinister than it likely was. Glenda pulled out an envelope to slide next to it, and he noted it was also done via a computer printer with no return address and postmarked locally.

She asked, "You knew about his liaisons with underage victims?"

Drayco nodded. "I saw some tapes that were made. But your father swears they were consensual encounters. And from what I saw on those tapes, there doesn't appear to be any coercion. Plus, the individuals looked borderline underage. For what it's worth."

"When I called him the other day, he said he didn't want to talk about it. Although he didn't deny it. Now, I'm in a quandary as to what to do."

"Which is precisely why the person blackmailing him was successful. Seal was afraid this would taint your cases and affect your career, so he went along with the blackmail."

"He didn't say anything about blackmail."

"The blackmailer seems to have targeted several people at Gaufrid Farm."

She sank into her desk chair and gripped the armrests. "When he went into that commune, I didn't understand. Made some comment about needing to get away from the rat race and reboot his life, but it didn't make sense. Then or now."

"Your father said it was partly due to his guilt over those encounters. And partly due to the fact he blamed himself for your mother's stroke, though he was told it was a genetic condition."

"The doctors said it was Moyamoya disease. I've been tested for it, myself."

"He also said if your mother hadn't passed away, they would likely have divorced since they'd drifted apart over the years. His affairs with the young women were a lapse in judgment, sure. But I gathered they were more of a way of seeking comfort and companionship."

Glenda rubbed her forehead. "I didn't know about all of that. Neither of them ever said anything. Guess I saw what I wanted to believe."

"He swears those tapes happened well before he went into the commune with no such encounters since."

She shook her head. "I should hand this info over to the appropriate law enforcement agency."

"Possibly. But have there been official charges against him from those young women?"

"None that I know of. I would have heard about them, trust me on that."

"If they were indeed consensual encounters, and there was no way for him to have known the girls were legal or not, doesn't that help your quandary?"

"I won't know until I pursue it further. Depends upon where the encounters took place, too. And how long ago. Did he say?"

"In the general Delmarva area. About twenty-five years ago."

"As you know, age of consent varies from one jurisdiction to another. Hell, even civil and criminal laws within the same state can conflict with each other."

"In Maryland, where some of these trysts occurred, it's as young as sixteen."

She grimaced. "Makes it hard on prosecutors."

"Not to prejudice your case, but you might like to know something I learned from another commune member. Your father has secretly given money to charities that fight abused and trafficked children. His way of making penance. And honoring your work and career."

Seal's daughter leaned back in her chair. "This is not going to be easy. No matter what I do or how it all ends."

"Have you gotten any blackmail notices, yourself?"

"No, nothing."

"Your work has some connection to one woman involved in my investigation. She may have started out her life in the U.S. as something of a semi-trafficked prostitute."

"I see a lot of that, unfortunately."

"This woman's name was Tasha Oleneva, who was brought to this country about thirty years ago as a teenager."

"Oleneva? Must be Russian." She sat down at her desk to check a database on her computer. "I've got numerous records about Russian and Eastern European women. Most exploited in some way, usually by rich and powerful men."

After tapping on the keyboard and scrolling through several files, Glenda said, "I'm not finding a Tasha Oleneva. Although names are often changed."

"She was likely brought here by a man named Stuart Wissler."

"Wissler." Glenda thought for a moment. "That name is familiar through one of my older cases." She tapped some more on the keyboard. "Yes, here it is. The woman involved in his case recanted, so charges were dropped. God, I hate those. These men have such an emotional hold over their victims."

"Benny Baskin's client met Tasha twenty-five years ago at a party at Congressman Curt Goldfeder's home."

"Goldfeder?" Glenda uttered a sound of disgust. "There were rumors about him for years. Thankfully, that man is dead now, but that means one less piece of evidence against this whole nasty business."

"Where one hydra's head is cut off, two more rise up."

"It's a never-ending battle. But I, for one, would be happy to be out of a job if it means no more abused women."

"You'd have plenty of other criminal cases to prosecute."

She said with a slight smile, "I see it keeps you in business."

He shrugged. "We're both in the justice business."

"It can be heartbreaking."

"But also rewarding."

She picked up a model plane. "My grandfather was a naval aviator during the second world war. Justice in his case was messy, bloody, and traumatic. He came back from the Pacific Theater a changed man. Not for the better. Despite having fought for the cause of decency and peace, the experience killed his spirit."

"You're afraid it will happen to you?"

"That a woman can't handle the stress, you mean?"

"Not at all. I wonder the same thing about me. All the time."

She put the little plane down. "You may be called as a witness if there's any court case concerning my father. But, after that's all over. . .we should go flying. It's easier to forget the cares of the world up there when you're above it all."

"I agree. And I'll be happy to take you up on your offer."

As Drayco left the courthouse, he was glad he'd met Seal's daughter, but disappointed he hadn't learned much. The anonymous letter was bizarre, too. Why now, after Lara's murder? And so soon after Drayco started asking the commune members questions, at that. Was it his newly bailed-out "goons" again?

Glenda Hettrick seemed like a good person, and she was certainly doing angel's work. Drayco was angry someone would make her collateral damage in this ugly game. There was already far too much preying on the innocent to go around.

Well, then. He was pretty good at games, himself, and it was time to try stacking the deck.

41

Sunday, October 4

The day dawned with a layer of marine fog, not so much pea soup as gossamer droplets. Fog had always fascinated Drayco, turning everyday objects into monsters or sinister shapes, something that sparked his imagination as a child. The fog was not an aviator's friend, however, but fortunately, Drayco was only going hangar-flying on this particular foggy morning.

He looked over at his companion. "I'm surprised Onweller let you escape Quantico jail. Usually, when you're in the middle of a thorny case, you even work on weekends."

"I told him it was related to my serial sexual assault case. Besides, it's Sunday. He can't expect me to work twenty-four-seven."

When Drayco gave him a stop-shitting-me look, Sarg grinned, "I'm cogitating on that case as we speak. In fact, I was cogitating the entire drive down. Sometimes you have to get out of the cauldron—"

"Before you wind up in the soup." Drayco hadn't heard that one in a long time, but it was one of his former partner's favorites.

"Besides, I owe you lunch, junior."

"That you do."

Sarg unlocked the car door. "You really think this stop will pay off?"

"One way to find out."

They climbed out of Drayco's Starfire and headed toward the green-roofed building with large letters across the orange-and-white-painted walls that spelled out *Stafford Airport.* As if the Cessna 172 and the Learjet parked nearby weren't enough clues. On any other day, it would be a treat to kick back in the lounge area and watch those planes coming and going.

But today, Drayco had another quest in mind. He and Sarg navigated toward the offices of Mouton Aviation, where Drayco asked to speak to the SkyBuzz FBO's owner, Fred Mouton. The man had jowls that would put a bulldog to shame as he rolled an unlit cigar around in his mouth.

Mouton greeted them with, "What's this all about?"

Drayco answered, "I spoke with someone on the Eastern Shore who's spotted a helicopter coming and going over the past several years. Either different ones or just the same one. I checked with the Coast Guard and Accomack and Salisbury Airports, but they didn't have any flight plans or knowledge of choppers that would fit."

"We do have some choppers based here. I could put you in touch with the owners. I own one, myself." He pointed to a blue-and-silver Bell 407 parked outside on the ramp.

"That would be helpful." Drayco looked around. "Do you have helicopter instruction, too? I've always wanted to add that rating."

"You fly?"

"When I can. Not often enough."

"That's what they all say," Mouton guffawed. "We do some instruction here. Even have a training sim. But what's so special about this particular aircraft?" He peered at Drayco. "You FAA? Did it bust the SFRA?"

"Nothing like that."

"Good, good. FBOs like mine here aren't happy to see the FAA stomping around."

"I can imagine. The helicopter I'm interested in may have a red stripe with a 'G' in the tail number."

Mouton scratched the back of his neck. "Don't like to give out information except to law enforcement."

Sarg pulled out his badge and showed it to him. Mouton said, "FBI, huh? No offense, but seeing the FBI is as bad as the FAA.

Unless said FBI are customers. We've got a few who fly out of here since Quantico's only a half-hour or so from here."

Drayco asked, "Would it be possible to check your records to see if a helicopter matching that partial description flew out of here the day of September fifteenth?"

Mouton moved to the counter to check the logs. "Looks like we had one that'd fit that description. Registered to an LLC. There's the full tail number." He turned the page around to show them and then scrolled through a few more logs. "Stopped here now and then to refuel. Flew in for the day and left that night."

"Do you recall the pilot and any passengers?"

Mouton chuckled. "Unless they're movie stars or some other mucky-muck, I don't recall all the transients who come and go. My brain's way too old for that."

On a hunch, Drayco described Adrik Gorky and Anikey Petrov, the two Russians he'd tumbled with. Mouton nodded. "I think I've seen the second one. More than once. Not an easy sort to miss, is he? Not all that friendly, either, which is the other thing I noticed. I mean, pilots have this camaraderie you don't find many places. We stick together."

"So you have no definite recall of these two men on that date?"

Mouton glanced at a wall calendar. "That was a Friday, wasn't it? I usually take Fridays off so I can have a three-day weekend. The guy who was on duty that day is on vacation right now. But you can ask him when he gets back in a week."

Drayco replied, "If it can wait that long. Do you have his contact info?"

"It's that serious, ay?"

"Potentially, yes. A murder case."

"Whoo, boy. Well, sure, David left me the numbers where he'll be staying. Got a condo share out on Grand Bahama. Lucky stiff."

Mouton rooted around in a drawer for some paper and wrote down the employee's name and phone number. When he handed it over to Drayco, he said, "You ever get serious about that chopper rating, you come see us."

"I live up in D.C."

"It's not all that far. You could come down on weekends and avoid all that peak traffic."

"Thanks," and Drayco took the business card Mouton handed over.

"Give you a teensy discount, too. Seeing as how you're FBI." He squinted at Sarg. "You fly, too?"

Sarg huffed. "I'm a landlubber. Though sometimes, this guy convinces me to fly with him. I'll never know why. Actually, I do know. The hundred-dollar hamburger."

"That'll do it. You ought to try Tangier, if you haven't already. Good views. Good seafood. Good people."

Back in the parking lot, Sarg waited for a Gulfstream jet to fly overhead before he asked, "So, your goon squad *may* have flown out of here in a chopper. To the Eastern Shore and back. Possibly. Not much concrete evidence to go on, there."

"It means someone could have flown between the commune and Stafford. I found an orange ribbon tied to a wooden fence post near a cleared area about a ten-minute walk from there."

"Like a windsock for a landing site?"

"Potentially."

"'Maybe' and 'potentially.' Not an investigator's best friends."

"Still. Gordon Aronson said there was a virus going around the weekend of the murder. Commune members were isolated in their cabins for the most part. Therefore easier for someone to sneak out, hitch a ride aboard Air Goon to Stafford."

"Then what? Take a car to McLean, do the deed, and reverse the process?"

"Timing would be critical. But I don't know of any other way to get from the Gaufrid Farm to McLean and back in one day."

Sarg shook his head. "Glad I'm just slumming with you on this one, pal. You always did love those bizarre puzzles."

ஐ ஐ ஐ

"Just like I remembered." Sarg took a deep breath and inhaled the aroma of garlic-fried potatoes and chili-onion brats. "Hope you approve. When you said I owed you lunch, I naturally thought of this place. Plus, it's on the way back from Stafford."

"Can't remember how long it's been."

"For you? I don't know. Me? Well. . .I don't know that, either."

"You haven't eaten here without me?"

"Guess not. Shame, too. You pianist types class up the place."

Beastie Brats & Burgers hadn't changed much from the days Drayco and Sarg used to snag lunch during Drayco's FBI tenure. Same red roof and big yellow sign. Same cook, Gabe Lanfear, who looked like he'd stepped out of the 1800s, with a handlebar mustache and slicked-back hair parted in the middle.

He saw the duo coming and reached over the counter to shake their hands. "Long time, no see, G-men. Budget cuts getting to you? Must be pretty bad. I mean, we're not the Waldorf Astoria."

"Better than," Sarg said, as he placed his usual order for a bratwurst with Swiss, sauerkraut, and extra mustard.

Drayco ordered the same thing, and they headed outside to claim a picnic table. Not that it mattered since they were the only ones outside.

Sarg slid onto a bench, and as Drayco joined him, Sarg asked, "You said you'd asked Sheriff Sailor to coordinate a court order and some dredging gear?"

"He's working on it as we speak. Going to be a lot of 'fun.'"

"Yeah, funsics, for sure."

"Thanks for your efforts in tracking down a certain former employer of our goons, by the way."

Sarg nodded. "Good hunch of yours, as it turns out. Had to pull a few teeth and call in a coupla favors. But it paid off."

As they waited for their food, Sarg drew an imaginary map on the top of the table. "Okay. If someone flew in a helicopter from the Eastern Shore to Stafford—"

"A flight would only take about ninety minutes."

"Okay, and then they made it to McLean in another ninety minutes. They climbed over the fence in Harry's yard, leaving blood behind on that tree. But how did they know when to go? That is when the victim would be there. The audio bugs?"

"Remember that Lara told her neighbor when she was going to see Harry."

"Our friends, the goons again?"

"I suspect so. It's possible that—" Drayco paused when he heard the sudden roar of a car coming close and fast. When he looked toward the sound, he immediately leaped over the table and pushed Sarg to the ground.

A split second later, someone from the car shot at them, the bullets pinging off the table close to where they were sitting moments before. The car didn't stop but kept right on going with a screeching of tires as it turned down a side road up ahead.

Sarg pulled himself up first and then extended his hand to Drayco to help him stand. "Worth going after them?"

"They have too much lead time. There are dozens of roads they could have taken around here. Besides, we should check on Gabe to make sure he's okay."

"Point taken."

Drayco wiped the grass and dirt off his shirt as both men rushed inside the diner. They didn't have to go far because a healthy-looking Gabe Lanfear greeted them and grabbed onto Sarg. "You guys okay? Was that gunfire I heard? 'Cause it sure didn't sound like an old clunker's backfire."

Sarg grinned, "Guess somebody didn't like my outfit today."

Gabe wagged a finger. "Trouble must follow you two around." Then he peered toward the front window. "Ordinarily, I'd say we should call the police. But you are the police. Kind of."

Sarg replied, "We'll take care of it. But stay inside, for now."

Relieved to see Gabe was okay, Drayco asked Sarg, "Get any ID on the driver or gunman? Or the car?"

"Nah, too focused on the blurry Drayco-missile headed my way. You?"

"A black, late-model sedan, which describes thousands of cars out there." After not getting the full plate of the truck that sideswiped him near Gaufrid Farm and now missing this car's plate, Drayco questioned his observation skills.

Sarg seemed to be reading his mind. "We're a great pair of detectives, aren't we?" But then, Sarg smiled. "Kind of like the good old days and the adrenaline rush. Have I told you lately how much I miss that?"

"Yes, you have. Several times."

While Sarg used his phone to notify his office, Drayco pulled out his own cellphone to alert the local police, followed by calls to the Fairfax PD and Benny Baskin.

As they waited for the police, Drayco glanced out the window at his car and moaned. "Oh, that's just terrific."

"What?" Sarg looked over and cringed. "Ooh, that'll set you back."

Drayco sighed when he saw the bullet holes in the Starfire. Only a couple, but the car would have to go into the shop and be expensive to repair. Now he wished he'd brought his Generic Silver Camry instead. He pushed that thought aside as he kept a close eye on the road in front of the diner in case the attackers returned.

Gabe carried out their drinks as they waited. "Guess that food'll be to go now. But you don't have to die of thirst in the meantime."

When the man disappeared in the back again, Drayco said, "If Onweller finds out you were involved in another melee involving local police—"

"Eh, whatever. He loves me. Mostly because he thinks I can be the bait to lure you back."

"He's going to be waiting a long time." Drayco grabbed a bottle opener from the counter for his soda. "You mentioned our goon friends, Petrov and Gorky. Since they were released on bail, they might be involved. And if they think *we* think they're involved—"

"They could be heading for the border. I'll check in with the office. Can't keep Onweller out of the loop now. Especially if your crazy theory about the murderer pays off."

Sarg swiped the bottle opener from Drayco, chugged some of his root beer, and belched. "A snake in your car, black widow spiders in your townhome, and now this. You're a popular boy. What's next, a ricin-tipped umbrella?"

"Funny you should say that. I joked about that very thing to Brock."

"And how did dear old Dad take it?"

"Not well."

Sarg didn't match Drayco's grin. "He's right to worry."

"These seem more like warnings than actual attempts to kill me."

"Then they don't know you very well."

Once the police arrived, the two men gave them what little information they had. After the cops went off to question Gabe, Drayco mused over this latest "warning." He and Sarg must have missed someone tailing them, which meant they really were two failed detectives. Unless. . . He went outside to examine his car and wriggled his body underneath to inspect the frame and undercarriage.

Sarg's head poked in from the side. "Whatcha doin'?"

Drayco grabbed a handkerchief from his pocket and tugged at a small black object that he pulled off and handed out to Sarg. When Drayco was upright again, they studied the device, and Sarg said, "GPS tracker."

"First a bug, now a tracking device. We can try to get prints, but that didn't work with the bug or my office."

Sarg was right about one thing. If the attackers thought this would warn Drayco away from the case—after putting the life of a friend of his in danger—they really didn't know him very well at all

42

Monday, October 5

Drayco hadn't slept much, an old and familiar refrain. But it wasn't getting shot at that kept him awake. He'd lain in bed for a while thinking about Harry's case, then when he got tired of lying down staring at the ceiling, he'd moved to the piano to play some Bach. But even that got interrupted when he thought he heard some scratching at his back door. He'd opened it, expecting Cat, but nothing.

Okay, so maybe being a human target-dummy had rattled his nerves. Or it was the convoluted case setting him on edge. He wasn't about to indulge in his cousin's premonition nonsense. He had a theory, he had a suspect in mind, and he might even be getting concrete evidence soon. That should be enough, shouldn't it?

He roused himself from his sleepless stupor to get dressed and meet with a certain pint-sized attorney. The first thing Benny Baskin said when Drayco strolled through his office door carrying a bag of Hava Java S'mores doughnuts was "'Bout time," as he grabbed the bag.

"Your wife still making you eat healthy?"

"Breakfast was an egg white omelet with spinach, mushrooms, and vegan cheese." Benny shuddered and started munching on a doughnut with a blissful expression on his face.

"I have to keep bribing you so you'll stay on my good side."

"Mission accomplished."

Drayco took his second-favorite seat since the Sangria-colored leather chair was AWOL again, as if often was whenever Benny knew Drayco was coming. Where did the man hide it?

Drayco asked, "What's the latest legal status with Alisa Saber?"

"She isn't being charged, so I guess Harry can continue to warm that jail cell solo. The cops found some fellow students of hers who swear she was in class at that time. Despite the prof's loosey-goosey records."

"And Harry's charges?"

"All depends on you. The Fairfax PD have no reason to drop charges yet."

"Not even after the shooting yesterday afternoon?"

"About that. You said you were on the way back from Stafford Airport with Sarg? Wanna fill me in?"

"I might have found a link between the goons and the Eastern Shore."

"Oh? Intriguing. But I need more than 'might' to take into court."

"Working on it, Benny. Trust me."

Drayco tried not to be obvious as he looked around for any sign of Nelia Tyler, but Benny noticed. "If you're wondering about my lovely and talented assistant, Nelia, she's got an important research project on deadline."

"I worry about that woman. I think there's a photo of her next to the word 'driven' in the dictionary."

"Yep." Benny licked his fingers to get the last of the doughnut icing. "By the way, I know about her impending divorce."

Drayco crossed his fingers behind his back. "You don't say."

"She's tough, but it's gotta be hard. Coming so soon after her parents' breakup."

"I can imagine."

"Nelia is onc of the sharpest people I've ever met. And frighteningly competent. Except when it comes to her personal life, and then she's a grade-A moron."

That made Drayco sit up straight. "What do you mean?"

Benny gave Drayco a sharp look before he handed over a fat folder filled with papers threatening to spill out. "Take a look at those."

Drayco flipped through the documents, which included some he'd needed an attorney's request to acquire. "Whoa. Looks like someone else in the commune had a tie to Stuart Wissler other than the victim."

"Who?"

"Twenty-plus years ago, Max McCaffin was a listed partner with Wissler in an investment scheme. And then Gordon Aronson about fifteen years ago."

"Welly well. Even more intriguing."

"Is that whole damn commune involved with this? Was it a front all along for some backwoods criminal enterprise?"

"Go find out, boy-o." Benny popped the last piece of another doughnut in his mouth and then added, "Oh, and that folder is yours. I made copies."

Drayco left the attorney sitting at his desk munching on his third doughnut and hoped Benny's wife would never discover Drayco was his supplier. Hell hath no fury than a wife's husband-slimming efforts being sabotaged.

On his way back to his car, he got a call from Alisa's newfound cousin, Olga Whitman. "I spoke with my Russian relative again—I think I told you about her. Tasha mailed her a letter twenty-five years ago, and she kept it. She wrote she'd had a baby and was happy and yet terrified. Knew she had to give her up but didn't know how. Didn't say much else, just that she was afraid for herself and for the baby. Wanted her to have a chance at a normal, happy life."

"Sounds like Tasha was involved in a bad situation and didn't know where to turn."

"Yes, that's how it seemed to me. I haven't told Alisa. Too painful. But I guess I thought you might want to know."

~ ~ ~

Drayco stood on the mansion's doorstep, surprised to see the fountains out front were still flowing. When he'd called Alisa to arrange an in-person meeting after he left Benny's office, she'd told him to come to Harry's house. That could only mean one thing—Darcie must be there, too, which also meant more future stepmother-daughter drama he wasn't really in the mood for.

Alisa was the one who let him in, although Darcie breezed into the room shortly afterward. She pointed to the floor around the bookcase. "Look, it's clean."

"You've had the crime-scene cleanup crew here."

"Before this, I had no idea such companies existed. But boy, am I glad they do. After they were through with the place, it feels normal again. Peaceful. Besides, I'm positive you'll get Harry free, and Harry will need to come back to a clean house. Alisa's been helping me."

Drayco looked over at Alisa, who added, "What she said. My father. . .Harry shouldn't have to see all of that when he returns."

Darcie grabbed Drayco's arm and led him to a sofa, patting the seat next to her. "Surely you know who did it by now?"

"I have a pretty good idea. But I need proof Benny Baskin can take to the police."

"But he's innocent. Why can't they see that?"

"They're doing their job, Darcie. They want justice, too."

Alisa spoke up, "Am I still a suspect? Is that why you wanted to see me?"

"No, I think you've been mostly dropped from the police's suspect list. Actually, I had a call from your cousin."

"Is she okay? Has something happened?" Alisa's face paled, and Drayco realized this cousin was one of the young woman's few remaining family members.

"She's fine. She spoke with her relative in Russia. Apparently, the cousin found a note sent to her after Tasha moved to the U.S. She'd had a baby and didn't want to give her up but knew she must, to keep her safe and give her a chance at a normal, happy life."

"And that baby was me."

"I think there's a good chance, yes. Tasha had gotten herself involved with the sex trade but not willingly."

"So where is she now? I want to see her, to talk to her."

"Unfortunately, I haven't found a trace of her yet. Rest assured, I'm not giving up."

Alisa got tears in her eyes and wiped them away with her sleeve. Darcie hopped up from her chair and went over to give the younger woman a hug. "Scott will find her. If anyone can, he will."

Deciding it was best to let the two women continue with their "bonding," Drayco told them he had an appointment. And he did, just not immediately. Darcie walked him to the door, and he said to her, in parting, "You've really matured since I first met you."

"You had a lot to do with that. Made me think more deeply about things. Helped me find my potential. You believed in me."

He left her and the house, musing on how he was a conduit for everyone else's happiness and success, yay rah. But a part of him was glad Darcie seemed truly content for the first time since he'd met her. As Harry said during one of Drayco's visits at the jail, "I have no delusions about Darcie being in love with me. But in time, I hope she'll come around." Maybe Drayco had just seen the initial signs of that.

He barely made it down the steps to the fountain when his phone rang again, making him almost hurl it into the water. But this time, it was Sheriff Sailor, saying that he'd run the plate of the truck Drayco saw at the commune that the beer supply company didn't know anything about. "It was registered to a subsidiary of Theunissen Trading. "

"Ah. I see."

Sailor asked, "Kind of seals the deal for you, doesn't it?"

"Unfortunately. Thanks for your hard work on this."

"No problem." Sailor paused. "Regarding those items you asked me about, I've got most of them lined up."

"Thanks, Sheriff. I'll see if I can get some hard evidence to make it worth your trouble."

"Be careful, Drayco. I'll be standing by."

Drayco hung up with the sheriff and considered his next move. The plan he had in mind was risky, but he hoped he could pull it off.

43

It took him five hours to get to the Eastern Shore, but Drayco made it to Gaufrid Farm just as twilight was ending. Gordon Aronson greeted him in the office with, "I was surprised by your request. But I've put some towels and things in Lara's old cabin where you can bunk tonight. It's unlocked."

"Thanks. I thought it would help to get a better feel for Lara's life here."

"Getting into the head of the victim?"

Drayco smiled. "In a manner of speaking." That wasn't his real reason for being here, but his answer seemed to satisfy Aronson. He asked, "By the way, Mr. Aronson, how's that trading of yours going. Make any big deals lately?"

Aronson blinked at him. "I don't. . .that is to say. . .I mean. . .deals?"

"You're still an active trader, I hear. I was wondering how your stocks are doing."

The director's face turned red as he stared at Drayco with a clenched jaw, not saying anything. After several tense moments, he replied in a voice even more hoarse than usual, "I think I mentioned how close the farm is to going bankrupt. I called a friend and set up an account again."

"How recently was this?"

"Last year. We need the money, desperately, Mr. Drayco. It was the only thing left I could think of."

As someone who also had a nearly bankrupt project on his hands, Drayco was sympathetic. But it didn't change the hard truth that

Aronson had been dishonest. "You haven't told the remaining members?"

"About the bankruptcy or the trading?"

"Both."

"Not yet. I was hoping to pull out a last-minute miracle." He looked warily at Drayco. "Are you going to. . .?"

"Not unless it becomes my business to do so." Drayco folded his arms across his chest. "I also found out you had previous dealings with Stuart Wissler. Is that why you were so quick to accept him as Lara's reference for membership?"

Aronson nodded. "It was not a wise partnership. In more ways than one."

"I understand you lost a lot of money."

Aronson uttered a humorless laugh. "Apparently, holding onto money is not my strong suit." He rubbed his forehead and said, "You know the way to the cabin. And now, if you'll excuse me, I think I need a stiff drink."

On the path to Lara's cabin, Drayco bumped into Seal, who seemed surprised to see him. But he also didn't appear as angry and antagonistic as he had over the phone when he accused Drayco of "ratting him out" to his daughter.

The man said, "You're here late."

"I'm staying overnight. In Lara's cabin."

"Scene of the crime?"

"Not exactly. But close enough."

Seal stood there awkwardly for a moment. "I want to apologize. For the other day. I've spoken with Glenda, and she explained everything. I offered myself up for prosecution, so she could keep her record clear. But she's not sure it will come to that. I gave her a couple of the names of the girls I could recall. She checked with them, and they swore it was consensual and knew about the tapes. They even admitted to watching them with me at the time. And lying about their ages."

"I guess in this day and age of kids sexting each other on their phones, it might seem quaint."

Seal sighed. "Yeah. I suppose."

"What are you going to do, then? Stay here?"

"If I don't end up in prison, yeah. I feel safer here, and Iguess I'll stay as long as it's open. After that, I don't know. Get some therapy. Not just for the sex, but for the guilt. The girls, my late wife, my estrangement with Glenda. . ."

Drayco nodded. "Any plans to see your daughter?"

"We talked about it. I'd like to. It'll depend on her and what she wants. I'm not going to push her. But since I just found out she's engaged, I want to meet my future son-in-law."

Drayco looked over toward Catherine's cabin down the way. "Is Catherine okay?" Her cabin is dark."

"Oh, that." Seal chuckled. "I think she's staying at Gordon's tonight. She and Gordon have been getting a little closer. Gordon told me you had a hand in that."

"Me? How?"

"Said you'd opened his eyes to what was right in front of him all along."

"Ah. I don't usually get to play matchmaker. Quite the opposite."

"It's good to see both of them happier. Don't know if it'll last. Who knows about any of us? Or about any relationships?" Seal waved at Drayco, "You know where my cabin is. Give me a shout if you need anything."

Drayco watched him head into his cabin and then checked into Lara's former abode, which was unlocked, as Gordon had stated. It looked the same as it had a few weeks ago when the director first showed it to him. Quiet, dark, and empty.

Its previous occupant was never coming back, and even her ghost had probably disappeared into the great beyond or the Hall of the Akashic Records. Or whatever reality followed this one, if any.

After sitting down on the couch, he put his feet up on a coffee table made from a vintage whiskey barrel. He checked to make sure the mini-recorder he'd brought was still in his shirt pocket and working normally. Then he fingered his necklace, which hid another device he'd affixed to it—a device he prayed he would never need. Last, but not least, he patted his pants pockets, with a flashlight in one, a pistol in the other.

With that, he leaned back and tried to relax. His task was better accomplished under the cloak of darkness, and he still had several hours until showtime.

44

Shortly after midnight, Drayco made his way to his first target, the swamps toward the edge of the property. He used a portable echo sounder borrowed from Dennis Frischman to determine the water's depth. . .about ten to twelve feet at its deepest point. Even during the day, the water was pretty dark and silty, making it hard to see much. At night, it was a black pool.

Then he headed for the commune's brewery, but unlike Lara's cabin, this building was locked. His pick-kit made quick work of that. Once inside, he took a look around using his pocket flashlight. Spying a leather-bound ledger on a desk, he donned some nitrile gloves and flipped through the ledger page by page, taking photos with his cellphone camera.

At first, the log appeared to be typical logs of ingredients and costs, brewing times, and distribution and shipping records. But after a closer look, he noticed some odd notations, spaced about every six to eight months—usually a cryptic number followed by "DES."

He slipped the cellphone back into his pants pocket and went over to examine the large barrels at the end of the building, giving them a good sniff and tapping off a tiny bit of liquid into a nearby paper cup. After propping his flashlight on the top of one of the barrels, he dipped his pinkie into the liquid for a quick taste—seemed to be normal ale.

But these looked like traditional wooden beer barrels. Just beyond those barrels, a shinier object peeked from behind a stack of wooden crates. When he got a closer look, it was a silver barrel like the ones he'd seen in the back of the Havisham Homebrew truck.

He tapped the side, verifying it didn't sound at all like wood. More like a metallic material, probably aluminum. What he hadn't seen on the barrels in the truck was clear on this one—more "DES" letters stenciled on the side. He said aloud to himself, "DES. Davos Electroplating Services?"

He bent over to rub the lettering, which was professionally done and didn't smell like recent paint. As he straightened up, a voice said quietly behind him, "When I learned you were staying overnight, I had a feeling I'd find you here. Then I heard you leave your cabin."

Drayco turned around to see Max McCaffin. Even in the dim light from Drayco's flashlight, he could tell Max was holding a Sig Sauer, presumably loaded.

As Max flipped on a light switch, Drayco replied, "Your tour of the brewery on my earlier trip was too brief. Thought I'd like to get a more thorough look around."

"With a flashlight?" Max chuckled. "I think we both know why you're here. I keep in touch with Devon Monk, and he told me you'd been out to visit him. Didn't say why. But I knew you'd eventually put two and two together. All our little warnings did nothing to scare you away. Pity."

Drayco spoke a little louder, hoping his pocket recording device was doing its job. "How long have you been dumping toxic waste from your old employer here?"

"Off and on. Believe it or not, I did try to make an effort to reform at first."

"What changed your mind?"

"A few years after I got here, some of my former colleagues-in-crime came to me with a plea for help. They needed to dump more waste, and the laws were getting too onerous to work around. So I decided I did miss some of my old life. What harm could it do? Just a bit of waste here and there since swamps and marshes make a perfect hiding place. Sometimes the feds get too close to my contacts at the plant, and we have to cool it for a while."

"You also had a leaking barrel once." It wasn't the best tactic he could come up with, but Drayco just needed Max to keep talking.

"What? Oh, yes, a small one. The neighbors complained, and I got a little panicked. But the ole 'swamp gas' excuse won out."

When Max moved a step closer, Drayco was actually relieved since he wanted to make sure the recorder got everything—no matter what happened. "You killed Ivon when he found out about the dumping? Or did you have Adrik Gorky and Anikey Petrov do it for you?"

The corners of Max's lips turned up into a mocking smile. "I offered Ivon a cut, but he didn't want to play ball. It's unfortunate."

"Lara Davidenko, too?"

"She was going to blab to Harry Dickerman about Ivon. Oh, she was perfectly happy to be part of the whole toxic waste scheme, as long as she got her cut. But I guess murder was too much for her delicate sensibilities."

Drayco baited the man some more. "I visited your mother at the nursing home. You weren't traveling to see her every month as you told everyone."

Max seemed surprisingly defensive. "She was taken care of. I saw to that."

"But instead, you were making 'business' trips on your weekends?"

"And a stay at my secret mansion up on Long Island. Registered in an LLC."

"Long Island? That explains a lot. Since you took the local bus only as far as Wilmington, I'd guess Petrov flew you on his helicopter the rest of the way. The same helicopter you used to fly to Stafford—before your goons drove you to McLean to take care of your 'little problem?'"

Max frowned, but his gun hand stayed steady. "You're very good, aren't you? I suppose I'd gotten too careless about all those pesky details."

"Fifteen plus years of waste in the wetlands here is a pretty big deal, hardly a trivial matter. Not to the environment, nor the people who live around here, or fish that swim in tainted waters."

"It's a polluted world, Mr. Drayco. We're eating it, drinking it, bathing in it, wearing it, breathing it, driving it. You can't escape, so you may as well profit from it, I say."

"What about Tasha Oleneva? Did you kill her too? All traces of her vanished around the same time you joined here. And you had ties to Stuart Wissler, her former handler."

Max's eyes widened, and he froze for a split second. "Oh, you really are good. Bravo. Tasha was another unfortunate problem. She tried to escape from Stuart Wissler's clutches. Stuart wanted to teach her a lesson and tasked me with doing it. But. . .it went a little too far."

Drayco tried a stab in the dark. "And she's buried in a barrel in those same marshes. Like Ivon Leddon."

"Why don't you find out yourself?" Max used his gun hand to point toward the large silver barrel in the back. "Climb in."

Drayco took a deep breath, trying to quell the sudden feeling of panic at the thought of being sealed inside the barrel. He *might* be able to survive a gunshot and escape—but even if he did survive the first attempt, Max could just shoot him again.

He'd counted on the fact Max might discover him, but he hadn't counted on Max owning a gun—a potentially fatal oversight. Drayco was a pretty quick draw, but he doubted he'd have time to use his own gun from his pocket.

As if reading his mind, Max said, "And do keep your hands where I can see them."

Drayco said slowly, "You're giving me the option of death by gun or death by suffocation, is that it?"

"Isn't it nice to have a choice as to how one meets their Maker."

"Why don't you just shoot me and then put me in the barrel? I'm assuming that's what happened to Ivon."

"Too much noise. Gordon and the others might not buy the 'night hunter' excuse again like they did with Ivon. This is much simpler. For me, that is."

Drayco turned toward the barrel Max had pointed out, noting one of the ends was open. "Is that my ride?"

"Yes. And quit stalling. It won't help you."

After taking a few more calming breaths, Drayco headed for the barrel and peered inside. Barely big enough for him if he folded his legs. Every fiber of his being was screaming at him to rush Max and to hell with the bullets. But what real choice did he have?

As he turned to face the barrel with his back toward Max, Drayco pressed the little emergency transmitter on his necklace. Then, with one last backward glance at Max, the gun, and the world outside, he climbed into the barrel.

He thought he was mentally prepared for the moment the lights went out as Max sealed up the top, but his racing pulse and full-blown panic attack had other ideas. All his best efforts to slow his heart rate failed when he heard the muffled noise of the forklift starting up.

Max wasn't trying to be gentle, and Drayco hit his head on the barrel as it was rolled around onto the forklift. He didn't dare waste any oxygen to rub his head or reposition himself, and each bump over the ground added a new bruise to his growing collection.

He tried to slow his pulse rate again, realizing he had a chance of being found since both Sheriff Sailor and Sarg knew where he was. But then, they expected him to be above ground, not below it. Did the device on his necklace have time to work? Or had he pressed it too late?

It was pitch black, but he could move his arms and legs. . .just not much. He reached out to touch the smooth surface of his prison. Maybe Ivon Leddon and Tasha Oleneva were also alive when they were treated to Max's little roller coaster of death. Hopefully, they lost consciousness first, if so. But Drayco was still aware of his surroundings, so that must be a good sign. So far.

He started playing the Moonlight Sonata in his mind, focusing on each note and imagining himself back at his piano. When he'd "played" through that, he moved onto another of Beethoven's work, the *Pathétique*, taking his time, making it all the way to the end. How long had it been now? How much longer would his air hold out?

Despite his best efforts, a tidal wave of panic arose in his brain again, and he fought the urge to thrash and kick and punch until he was free. Try as he might, he couldn't push aside the memories from his childhood that were boosting his claustrophobia into extreme high gear.

He hadn't told Sarg the whole story about his cousins stowing him in that footlocker—they thought it would be fun to bind and gag him like in the movies. Then they went off to play football, leaving Drayco wedged in the locker alone for hours in the dark, without any light or sound.

He forced himself back to the present to focus on the physical sensations of the barrel. The pain of the bruises and his headache was a welcome relief, and he used the pain to concentrate his thoughts on those sensations.

But he began to notice something he'd never experienced before, not even in the locker—an almost complete synesthesia void. No sounds, no colors, no shapes, no textures. That was the most unnerving thing of all. As his heartbeat got louder and louder, it became a growing, mossy, reddish-brown mass rising to engulf him as he became more aware of it.

When the panic threatened to swell to a new peak, he slowed his breathing to conserve air like he used to do during his freediving trips on Bonaire. Because he knew exactly where he was—in the same marshes and swamps along with dozens of barrels of toxic waste, thrown there to join Ivon and Tasha.

The dreams he'd had since the summer after the suicide-sonata case came back to him, too. Dreams of falling into a deep pit with water and nearly suffocating. Believers in the Akashic Records would say he had glimpses into his own future. . .and even his own death.

He also recalled Gordon Aronson's words when the man wondered if all of his efforts and his dreams had been worthwhile. What about Drayco's efforts and dreams? His lost piano career, his abbreviated FBI years, all his cases and the victims he'd tried to find justice for—had it all been worthwhile?

Then, a sense of peace and calm settled over him. Maybe this truly was the end. Suffocating wouldn't be his preferred method of leaving behind this vale of tears, but it could be worse. Or maybe he was having hypoxic delusions. Like the sudden image of his mother singing the same Scottish lullaby, "Baloo Baleerie," she sang to him when he was small. *Hush-a-by, hush-a-by, Go to sleep my babe, Watch o'er him, blessed angels, My babe will sleep*. Was she telling him now to let go?

His peace was shattered by a rough shaking and rolling that flung him around in the barrel, leaving behind a number of painful new bruises. Finally, the barrel came to a rest, and he thought he heard shouting and a scraping sound on top.

As the lid slid off, the lights overhead were so bright, he had to close his eyes. When he opened them again, he saw dredging gear and trucks and dark figures silhouetted against the lights.

Two of those figures peered into the barrel as the faces of both Sheriff Sailor and Sarg came into focus above him. He groaned. "You two together in the same place? I've died and gone to hell."

Sarg laughed and then said with relief in his voice, "Eh, he's fine."

45

Tuesday, October 6

Drayco hadn't slept any after being up all night with the EMTs and Sheriff Sailor and his crew. Once he'd been given the all-clear, he'd updated Benny Baskin, who was immediately going to press the Fairfax PD for Harry's release. Benny was helped by the fact the FBI caught Max McCaffin as he tried to skip the country via Norfolk International airport. It took most of the day to secure Harry's freedom, but Benny triumphed.

McCaffin had apparently used the commune car for his escape, but Drayco's Generic Silver Camry was also missing. A bulletin was sent out in case the goons were seen driving it. But it meant Drayco had to grab a rental car to head back to D.C. late in the afternoon.

On his return trip, Drayco received a call from Sheriff Sailor, who filled him in on the Gaufrid Farm excavations using the same dredging gear that had saved Drayco's sorry ass. They'd found other barrels, and not all of them contained toxic waste.

Grateful when he made the five-hour drive to the District safely despite being sleep-deprived, Drayco made a stop by his townhome to take a shower and grab clean clothes. Then he headed back down I-95 toward a place and a person who was not going to be glad to see him.

He parked in the monastery lot, and after a quick check with the front office, he found Daven Monk back in the cheesemaking shop. Monk was sitting bent over a mini-press when Drayco arrived. But one look at Drayco's face, and he immediately straightened up. "I'm guessing you didn't come back for more Gruyère."

"The first time I spoke with you, you said there were no secrets here. That you'd told the order—and me—everything."

Monk gave Drayco a razor-sharp stare and didn't reply.

"But there is one secret I'll bet you didn't pass along. Toxic waste dumping."

The other man's jaw moved back and forth, but he still stayed silent, so Drayco continued, "I saw a truck at Gaufrid Farm the other day. At first glance, it appeared to be from a beer-making supply company. Yet, when I called that business, they didn't know anything about it. Turns out, the plate was registered to the parent company of Davos Electroplating Services. A company Max McCaffin used to work for."

Monk again stayed silent but fiddled with the edges of the press.

"Funny thing, too. A couple of thugs broke into my office ten days ago. And I recently found out, thanks to a colleague, that those two men also once worked for Davos Electroplating Services."

Monk sighed and leaned back against the press. "I see."

"So, I took a wild guess that McCaffin's former company was still up to their old tricks and using the beer barrels and supplies as a cover. And since you were assistant brewmeister, you had to know about it."

"It's true. I knew."

"Why not reveal this to the monastery as part of your sins when you were admitted?"

"I would likely go to prison. And I can't do as much good in prison as I can here." Monk gazed briefly at the ceiling as if looking for a little divine intervention. "I'm paying for my sins in my own way, you see? I've wrestled with the knowledge since I arrived. Every single day. Perhaps that sounds insufficient. But I'm being honest."

"Did you receive money from the scheme?"

"Yes, but I gave it all to the monastery when I joined. For charity work. You may not believe me, but it's true. I really am penniless beyond what I make here." He paused and then added, "Are you going to tell the police?"

"I have to. The FBI arrested Max McCaffin trying to flee the country. After he tried to kill me."

Monk's eyes widened. "Dear God."

"Did you know he killed Ivon Leddon when he found out about Max's activities and threatened to expose him for it?"

Monk sat perfectly still. "No. No, I didn't. Or. . . I half-suspected and simply refused to see the truth. But, I guess I'm ready to deal with however this turns out. 'Therefore do not worry about tomorrow, for tomorrow will worry about itself.' Matthew chapter six, verse thirty-four." He added with a slight smile, "And there's always prison ministry, right?"

Drayco left him there, staring at the cheese vat in deep thought. Drayco just hoped he'd made the right decision to trust Monk not to flee, to stay and "deal with however this turns out." If the man did flee, then Drayco would never hear the last of it from the FBI, EPA, FTC, or even Detective Shephard King. Drayco considered himself a decent judge of character, but only time would tell if he'd got this one right.

~ ~ ~

By the time Drayco finished with Monk, he realized how exhausted he was. But he'd promised Darcie he'd stop by her fiancé's mansion for a mini-celebration. It took every ounce of energy he had left, but he made it just as twilight fell and moonlight started to cast shadows in the fountain.

Once Drayco was settled inside, he noted with amusement that Darcie kept filling Harry's coffee cup every time it got down even an inch below the rim. Harry shook his head. "I know I said I craved good coffee the whole time I was in jail, Darcie. But my kidneys will never be able to handle this deluge."

She kissed him on the cheek. "Just want you to feel welcome back in the land of the living."

"Believe me, I do. You've gone above and beyond, my love."

Drayco kept a close eye on Alisa Saber, who kept staring at Harry as if trying to convince herself he was real. Drayco had already told her they'd dug up the other barrels from the marsh behind Gaufrid Farm. Unfortunately, they found not only the toxic waste but a barrel with the remains of Ivon Leddon and another with what was likely Tasha

Oleneva. They'd have to wait for verification via a DNA test from material to be taken from both Alisa's new cousin and Alisa herself.

When he'd told Harry about the find, the man was as teary and subdued as Alisa, to his credit. Harry had only one night to fall in love with Tasha, but it was obvious he'd fallen hard. It was also heartbreaking to know Alisa never had the chance to meet her mother, and now she never would.

While Darcie was hovering over Harry, Drayco asked Alisa, "What are your plans now?"

"I've still got my studies. Especially since I'm switching gears."

"To what?"

"Poly-sci and then law school."

"Why the change?"

"I want to be able to help women like Tasha. Victims of trafficking, abuse, assault. I can do that better with a law degree than biology."

Drayco smiled. "Then I need to put you in touch with someone. Her name is Glenda Hettrick, and I think you and she will have a lot in common to talk about."

Alisa added, "I'm also going to study Russian. To feel closer to my mother." She cut off the gentle cautionary warning he started to give by saying, "I know in my heart the body they found is my mother's."

He nodded. "Have you spoken with your cousin in Maryland?"

"Yes, and we're thinking of making a pilgrimage to Russia in the not-too-distant future. To track her relatives and my mother's family. Whether they welcome it or not. I have to know."

Drayco thought back to his own recent encounter with his long-missing mother. Alisa didn't have to explain it to him—he understood her reasoning quite well. He grabbed a manila envelope he'd laid on a table and handed it to her.

"What's this?"

"It was found among Max McCaffin's possessions at Gaufrid Farm. The original copy of your mother's photo."

She took the envelope and smiled as she gave him a big hug. "Thank you. For everything."

Since he'd been there for over an hour and felt it best to let the new little family get acquainted, he told them he had to leave for a meeting with Benny Baskin.

"Are you sure you can't stay longer?" Harry stood up when Drayco did.

"Some other time."

Harry strolled over to pump Drayco's hand. "Can't thank you enough for all you've done. You've saved me and brought me back to my fiancée and a daughter I didn't know I had. I was wrong to doubt your talents. But I plan on making it up to you."

Drayco gave a brief smile. "No need. I'm just happy it worked out."

Darcie accompanied him to the front door where he asked, "Wedding plans still on?"

"Absolutely. Besides, that girl needs a Mom. Or a big sister."

"Why, Darcie, that's very mature of you."

"I'm thirty-five. Much easier to be an instant Mom than go through with that whole varicose vein, stretch-mark thing."

He blinked at her, but when he saw she was joking, he smiled. "I think you'll be an amazing mother-sister-wife."

She grabbed his hand and held onto it for a few moments. "Thank you, Scott. For everything."

As he left the trio and mansion behind, he didn't feel sad or relieved or happy or much of anything. He wasn't sure why. Since his ordeal in the barrel, he'd thought about his career and relationships and realized he'd made a lot of mistakes. But mostly, he was numb, devoid of any feeling. Brock had suggested he talk to Dr. Kinder, but Drayco wasn't in the mood for therapy.

He hoped his malaise hadn't affected the welcome-home party for Harry, but he wasn't lying about that appointment with Benny. He'd just lied about the time since it was tomorrow, not this evening. Hopefully, some sleep and two CCs of a wisecracking attorney in the morning would be the antidote he needed.

46

Wednesday, October 7

Not only had Drayco made it back to his townhome for much-needed sleep, he'd collapsed onto the bed fully clothed. He was groggily awakened by a call from Benny Baskin reminding him he'd agreed to pop by for updates. Drayco looked at the time. He'd slept thirteen hours straight. No wonder he was late for his appointment.

Benny was alone when Drayco arrived at Benny's office, and the attorney ushered Drayco into his favorite Sangria-colored leather chair with a snifter of brandy. "Looks like you could use a shot of something."

Drayco took the drink gratefully. "Sarg called this morning. They arrested a couple of bigwigs at Max's former company, who'd arranged the toxic waste dumping scheme. Davos Electroplating Services has been shut down."

"Excellent! Good to hear."

Drayco added, "Oh, and I asked Gordon Aronson about that investment scheme of Stuart Wissler's he got mixed up in. Aronson says he hadn't wanted to talk about it since because he was too embarrassed. Lost a good bit of money, I understand."

Drayco's expression must have been on the glum side because Benny added, "Speaking of money. Sorry to hear about your car."

He knew Benny wasn't referring to the bullet-riddled Starfire. The goons weren't the ones who stole Drayco's Generic Silver Camry as a getaway vehicle. Instead, Max had pushed it into the marsh so he could tell everyone Drayco left during the night due to an emergency. But when Max had second thoughts about hanging around, he took off in the commune car. Drayco was actually grateful to the two Russian goons for shooting up his Starfire since it was in the shop—keeping it from being the vehicle suffering such an ignominious end.

Benny bounced up and down on his platform-heeled shoes. "Ironic, isn't it?"

"What, the car?"

"No, Max McCaffin. Lara née Minna was killed because she went to Harry for help. It was part convenience and part vindictiveness to frame Harry. But if he'd simply made her death look like an accident, we'd never have been involved. He'd have gotten away with it. All of it."

Just then, Nelia Tyler entered the office and gave Drayco a shaky smile. "There you are. Fresh from the bowels of the swamp and looking normal."

"Normal-ish. But I'll take it."

She bit her lip. "At least Sarg got to meet Sheriff Sailor finally. They get along really well. And they've been comparing notes."

Drayco leaned over and banged his head on Benny's desk. "Kill me now."

Benny piped up, "Someone tried to do that and failed, as I recall."

"You should have had backup." Nelia's voice had a sharp edge to it, and she seemed uncertain whether to sit or stand as she kept rubbing her arm self-consciously.

"I did," Drayco replied with more force than he'd intended. "Sheriff Sailor was already lining up the dredging gear. And I'd told both Sailor and Sarg my plans about staying the night at the commune. Plus, I had the emergency ELT device."

Benny looked between Nelia and Drayco and said, "Will you look at the time. I promised my accountant I'd give him a call at eleven, and it's half-past." When Drayco started to get up, Benny added, "No, no need. I'll call him from down the hall. You kids get caught up. Share a few jokes at my expense."

After Benny had left, Nelia eased around to the side of the desk, still standing. She blurted out, "How's the concert prep going?"

"Don't know if it's ever going to happen. Or if I even want it to." He paused, then asked, "How's it going with Tim?"

"Not great at the moment." She managed a small smile. "He's still begging me to take him back, but I'm leaning toward not."

"Ah. Well, I'm sure you'll know the right thing to do."

She started rubbing her arm again. "I was pretty scared when I heard *after the fact,*" she emphasized the words with a frown, "about you being in that barrel. It's one thing to deal with personal danger, I'm used to that. But it's another when someone you lo—" She checked herself. "Someone you care about is in danger."

"But I'm okay."

"*Now*, you are." She blinked back a few tears. "Please understand. I can't get in another relationship. After all of this with Tim and my parents and work and law school. . .I need time. Time to breathe, time to find out who I am and what I want."

He nodded but didn't add anything. He did understand, truly. And he was willing to sacrifice his feelings for Nelia to give her whatever she needed to find her footing. He'd be waiting if she still felt the same way. Maybe she would, maybe she wouldn't, but as that old cliché said, sometimes you have to set love free. It may come back to you, but if it doesn't, it was never yours in the first place.

~ ~ ~

Someone slid into the seat next to Drayco, but he didn't look up until a familiar voice said, "Tried to call you, but I didn't get an answer. So I figured I knew where you were. You used to come here after hard cases, the kind that wring out your soul and make you want to drink yourself to oblivion. In fact, I think you almost belted me the last time we were here."

Drayco looked down at the shot glass in his hands. Part of him wanted to be alone, but at the same time, he welcomed Sarg's company, someone who'd known him longer than most people. Drayco lifted his glass by way of a tacit welcome.

Sarg ordered another round for both of them, then told Drayco, "Glad my colleagues snagged McCaffin before he could leave the country."

"I'll have to send them a fruit basket."

When Sarg snorted, Drayco added, "The Fairfax PD chased down the goons a second time. Got them to talk, and they admitted working for Max. And being the ones to send me the spiders and play shooting gallery with us as targets. They said Max was the one behind the snake. And also Ivon's murder, although they're all pointing fingers at each other for that one."

Sarg lifted up his mug in salute. "I'll have to send the Fairfax PD a fruit basket. Or maybe a doughnut basket."

Drayco took another sip of his whiskey. "Thanks again for your part in my rescue."

"Thank god it timed out in our favor. Your hunches are awfully risky, my friend. You cut this one too close."

"Guess it paid off."

"I'll say. That helicopter of Fred Mouton's is pretty sweet, by the way. He got me over to the commune in time to help with the rescue."

Drayco downed more of the whiskey, letting the acrid liquid trickle down his throat and waiting for the buzz to kick in.

Sarg looked at him out of the corner of his eye. "So. . .must have been tough in that barrel. The claustrophobia and all."

With a shrug, Drayco replied, "I knew you'd come through."

Sarg raised an eyebrow but let it go. "The commune will have to be shut down while the toxic waste investigation is ongoing."

"If we're taking bets, I'll bet it never reopens."

"What about the monk, Daven Monk? I hadn't heard the latest."

"Looks like he likely won't get any jail time." Drayco crossed his fingers in the air. "That's ultimately up to a judge or jury, but he's allowed to stay at the monastery since the goons swear he had nothing to do with any violence. Just that he knew about the toxic dumping."

"Good. That's good. I think."

They sipped a few moments in silence until Sarg asked, "Ever figure out that name on the adoption worker's piece of paper? Gerasim, wasn't it?"

"Since Tasha isn't here to tell us, I can't be sure. But the English equivalent of the Russian man's name of Gerasim is Harrold, or—"

"Harry."

"Perhaps in the confusion and raw emotion from having to give up her baby, she mixed up the two names, reverting back to her native Russian version."

Drayco couldn't get that image out of his mind—of the young woman far from home, trapped in a prostitution scheme, tearfully handing over the one good thing in her life, her baby girl.

Sarg patted him on the back. "Cheer up. I heard from Baskin that Harry Dickerman is giving the Opera House a big chunk of money so it can be completed. You okay with that?"

"I'm happy on one level, sure. But am I supposed to rename it after him, now?"

"That's usually what people do."

"Not that I'd name it after me. That was never on the table. I was thinking more along the lines of naming it after pianist Konstantina Klucze. Since her last concert was there."

"And it was her Chopin manuscript found there."

"Maybe I can mash 'em up. Dickerman-Klucze."

"That's a mouthful."

"Harry Konstantina is worse."

Sarg grinned. "Nice to see a bit of that sarcastic wit returning."

"I'm just not sure how to feel about any of this. Don't know why. It's hardly the same as after Mom's case, not even close."

"S'always hard after wrapping up a case. Any case."

Drayco sighed. "It's made me realize the shallow, plastic nature of life these days. The monks and their digital loudspeaker chants, the fake commune, the whole murder scheme just to hush up an environmental crime. Where's the sense of it all?" He shook his head. "All the personal losses, all the impermanence of life."

Sarg held out his hand as if wanting Drayco to shake, which Drayco did out of instinct. Sarg said, "You forgot about this. The human touch. This is what's real, my friend. Always has been, always will be."

Drayco mulled over his words and then nodded with a small smile.

Sarg added, "And you know I've got your back. You can call me any time, day or night."

Drayco stared at him, recalling those same words from Drayco to Nelia. Sarg looked at him quizzically. "What?"

"Nothing. It's just. . .thanks."

"*De nada.*"

They sat there drinking in silence, side-by-side, surrounded by the low murmurs of the other patrons, the buzzing of an overhead light on the fritz, and the smell from old beer stains on the carpet. Drayco didn't see many couples, just solo drunks or soon-to-be drunks. He started chuckling.

Sarg looked around. "What?"

When Drayco pointed out that they were one of the few "couples" in the place, Sarg replied with a mocking grin, "This isn't about that bromance thing again, is it?" which only made Drayco laugh harder.

Sarg clapped him on the back. "You did good, junior. Freed an innocent man and reunited him with his daughter, who also found closure about her mother. And you put a stop to some bad environmental chicanery. I'd say that's a pretty fine day's work."

"Maybe so." Drayco took a sip of his whiskey. "Maybe so." As an image of the Opera House popped into his head, he found himself planning out more of the renovations and a few new ideas he had. Troy Mehaffey was going to be one busy contractor. Drayco had also thought up another name for the place—the Maura McCune Performing Arts Center, named after Drayco's mother. Hopefully, Harry would approve.

Existence was forever a search for something. Love, fame, success, acceptance. The truth. Patching up holes left by the missing people in your life. Sometimes, you got it right and sometimes you didn't. What was it Beethoven once said? *I will seize fate by the throat; it shall certainly never wholly overcome me.*

Drayco raised his mug to clink against Sarg's in a toast to the successes and to continuing the search. Sarg said, "To a good day's work, then."

Drayco smiled as he added, "And to grabbing fate by the throat."

CPSIA information can be obtained
at www.ICGtesting.com
Printed in the USA
BVHW030545260721
612251BV00006B/55/J